P9-CQM-877

The Spiritual Doorway in the Brain

A Neurologist's Search for the God Experience

KEVIN NELSON, M.D.

DUTTON

DUTTON
Published by Penguin Group (USA) Inc.
375 Hudson Street, New York, New York 10014, U.S.A.
Penguin Group (Canada), 90 Eglinton Avenue East, Suite 700, Toronto, Ontario M4P 2Y3, Canada (a division of Pearson Penguin Canada Inc.); Penguin Books Ltd, 80 Strand, London WC2R 0RL, England; Penguin Ireland, 25 St Stephen's Green, Dublin 2, Ireland (a division of Penguin Books Ltd); Penguin Group (Australia), 250 Camberwell Road, Camberwell, Victoria 3124, Australia (a division of Pearson Australia Group Pty Ltd); Penguin Books India Pvt Ltd, 11 Community Centre, Panchsheel Park, New Delhi–110 017, India; Penguin Group (NZ), 67 Apollo Drive, Rosedale, North Shore 0632, New Zealand (a division of Pearson New Zealand Ltd); Penguin Books (South Africa) (Pty) Ltd, 24 Sturdee Avenue, Rosebank, Johannesburg 2196, South Africa

Penguin Books Ltd, Registered Offices: 80 Strand, London WC2R 0RL, England

Published by Dutton, a member of Penguin Group (USA) Inc.

First printing, January 2011
10 9 8 7 6 5 4 3 2 1

Image credits:
Images 3, 5, 6, 9, 10–12, 13a, 13b, 15a, 15b, 16, and 18. Courtesy the author; Image 1. Used with permission of Professor Charles Lieber; Image 2. Adopted with permission from Zeman, A. "Persistent Vegetative State." *Lancet* 350 (1997): 795–799; Image 4. Adopted with permission from Addis, D. R., M. Moscovitch, M. P. McAndrews. "Consequences of Hippocampal Damage Across the Autobiographical Memory Network in Left Temporal Lobe Epilepsy." *Brain* 130 (2007): 2327–2342; Image 7a and 7b: Damasio, H., T. Grabowski, R. Frank, A. M. Galaburda, A. R. Damasio. "The Return of Phineas Gage: Clues About the Brain from the Skull of a Famous Patient." Science 264 (1994): 1102–1105; Image 8. Gazzaniga, M. S., J. LeDoux *The Integrated Mind*. New York: Plenum Press, 1978; Image 14: Bradford Cannon Papers, 1923–2003, H MS c240. Harvard Medical Library, Francis A. Countway Library of Medicine Boston, Mass.; Image 17: Adopted with permission from Maquet, P., P. Ruby, A. Maudoux, G. Albouy, V. Sterpenich, T. Dang-Vu, M. Desseilles, M. Boly, F. Perrin, P. Peigneux, S. Laureys. "Human Cognition During REM Sleep and the Activity Profile Within Frontal and Parietal Cortices: A Reappraisal of Functional Neuroimaging Data." *Progress in Brain Research* 150 (2005): 219–227.

REGISTERED TRADEMARK—MARCA REGISTRADA

LIBRARY OF CONGRESS CATALOGING-IN-PUBLICATION DATA
Nelson, Kevin.
The spiritual doorway in the brain : a neurologist's search for the god experience / Kevin Nelson.
p. cm.
Includes bibliographical references and index.
ISBN 978-0-525-95188-9 (hardcover)
1. Psychology, Religious. 2. Spirituality—Physiological aspects. 3. Brain.
4. Neurosciences—Religious aspects. 5. Neuropsychology. I. Title.
BL53.N45 2011
200.1'9—dc22 2010037561

Printed in the United States of America
Set in Adobe Garamond Pro
Designed by Daniel Lagin

To my parents

Sine qua non

In loving memory of Dorothy Nelson,

Raymond F., and Barbara Nelson

CONTENTS

THE SPIRITUAL DOORWAY IN THE BRAIN

Prologue

AT THE FOOT OF THE BED

"Your adversary the devil, as a roaring lion, walketh about, seeking whom he may devour . . ."

—I Peter 5:8

This book began nearly thirty years ago, when I was training as a neurologist at the University of New Mexico Hospital in Albuquerque. Neurology interns do a year of internal medicine, and I had a Hispanic patient named Joe Hernandez, who was under my care for his diabetes and heart disease. Joe was a weathered man, a laborer who'd spent much of his life outside in the Southwest desert. We quickly formed a bond, although our backgrounds were completely different.

I had been raised in a modest but comfortable home in the small town of Grand Haven, a conservative Dutch Protestant community on the shore of Lake Michigan, nearly two hundred miles north of Chicago.

Even as an adolescent, I was fascinated by the study of the brain and knew that I wanted to be a neurologist—not a family doc, cardiologist, or any other medical specialty. At Michigan State, I gravitated toward the emerging field of behavioral neurology, which studies how the brain coordinates the amalgam of thoughts, feelings, perceptions, and memories we call the self. It investigated what happens when you stimulate or disable specific parts of the brain: why, for example, a stroke victim who suffered a lesion on the right side of his brain completely ignored the left side of his body; or why, as Oliver Sacks famously noted, a man could mistake his wife for a hat.

That was the kind of work I wanted to do. My undergraduate thesis involved chemically stimulating a tiny region of a female rat's brain and then recording whether she would let a male rat mount her. Like the behavioral neurologists, I wanted to understand how specific behaviors stem from precise locations in the brain.

Through localizing brain function, the behavioral neurologists had shown that who we think we are is a complicated and rather fragile synthetic process orchestrated by our brains. When something interferes with that process, our reality and sense of self quickly and dramatically fragment. While most of us view our "self" as concrete and coherent, akin to, say, Leonardo da Vinci's portrait of the Mona Lisa, to a neurologist the self is more like Picasso's cubist portrait of Dora Maar, his lover and muse: a fragmented amalgam of fractured planes. Or, if you prefer Impressionism, our view of the self is a little like a water lily by Monet: at a glance it looks coherent, but up close you realize its harmonious appearance is an illusion, that the object you saw at a distance is actually a bundle of discrete and unconnected parts.

It was this fragmentary nature of the self that I wanted to study, which is how I found myself as a neurology intern at the University of New Mexico Hospital, caring for Joe.

Not long after entering my care, Joe had a massive heart attack and spent more than a week in the ICU. Frankly, I didn't think he was going to make it, and I was much relieved when he pulled through.

I attributed his recovery to his tenacious spirit and sound medical care. But Joe had a very different explanation for why he was still alive, as I learned soon after his discharge when he came to see me for a follow-up examination.

"Doc, I have a gift for you!" he exclaimed right off and handed over a photo of an incredibly realistic oil painting, a self-portrait. Joe was in the ICU. Bright lights blazed. A battery of medical instruments stood to one side of his bed. Intravenous bottles hung above him; tubes fed into both his arms. Although it was clear that he hovered between life and death, Joe had depicted himself as awake and alert.

At the foot of his bed stood Satan, a devil with horns in a red robe.

"He had come to claim my soul," Joe explained. "But look, here is my guardian angel." An angel with a halo and wings outstretched stood between Satan and Joe.

"The devil, he was stronger," said Joe. "He was about to take my soul. And then my savior Jesus appeared and the devil vanished. I was greatly relieved, for then I knew my health and soul were safe, and I would be allowed to remain on earth a little while longer. Jesus came to save me. Doctor, it is a miracle!"

"Perhaps it was a dream," I suggested as gently as I could. I judged Joe's near-death experience to be a quaint blend of cultural myth and illusion. My training in neurology told me that he had been hallucinating. Joe, however, was adamant: his experience had been absolutely real.

I often thought about Joe's painting in the following months. What struck me was the vividness and intensity of the experience it portrayed—characteristics that sharply distinguished it from common illusions or dreams. Even at the beginning of my career, I knew that patients, in retrospect, typically recognize hallucinations for what they are: hallucinations. But Joe was firmly convinced that a battle for his soul had really occurred in the ICU.

I knew that the brain that fuses Monet's strokes of color to perceive a water lily was also responsible for the hyperrealistic image Joe saw when he was close to death. As a fledgling neurologist, it was natural for me to take the same tack as Picasso and deconstruct the brain processes battling for Joe's soul. What was the locus of brain activity responsible for what had clearly been for Joe a deeply important religious experience?

In the late 1970s, neurology had little or nothing to say on these points. Why was that? I wondered. I filed away these questions and practiced conventional neurology, focusing on diseases of muscles and nerves. But I came to realize over the years that the power of Joe's experience had made an indelible impression on me, and the questions it raised about the nature of brain activity near death persisted. I kept my ears tuned for other experiences like Joe's, and, to my amazement, I found these or similar experiences were common among neuroscientists and respected physicians in other fields, who were often left shaken by them, guarding them closely and divulging them somewhat reluctantly, as if they were kooky anomalies, shameful secrets that ran afoul of science. I began to collect these stories into what I thought in a vague way might one day become a book. And I continued to puzzle over why, aside from an occasional sporadic report, neuroscience dismissed near-death and what I had come to see as other related experiences: out-of-body events (which occur in one in twenty people), visions of dead relatives or spiritual teachers, and a pervasive sense of bliss that often has to do with feeling union with God or the universe: a feeling that has been called "oneness."

The way spirituality manifested itself in the brain was largely ignored—even derided—by my peers. There were reasons for this, of course, but the roots of neurology's derisive attitude toward the spiritual came about relatively recently, when neurology and psychiatry split into separate disciplines in the early twentieth century (people often forget that Freud began his medical training as a neurologist). Psychiatry took purview on subjective experience and *mind*, while neurology and neurologists focused exclusively on the *physical* brain.

The willful ignorance and offhand dismissal by neuroscience of such a significant part of human experience struck me as particularly egregious in the 1990s, the "Decade of the Brain," when the technology that allowed us to visualize the living brain was exploding beyond anything we could have hoped for or imagined. Functional MRI and positron-emission tomography (PET) scans permitted us, for the first time, to see how the brain worked and where it was active or inactive as it performed tasks such as speech, memory, complex thinking, body movement, sex, and dreaming. Our ability to "localize" brain function and regions important to subjective experience advanced by leaps and bounds.

The questions posed by Joe's painting were still with me: Was it possible to investigate the physical brain processes underlying near-death experiences and other spiritual events? Could we detect what parts of the brain were active in experiences of awe, religious visions, and what are sometimes called higher or altered states of consciousness? What was the neurological basis of religious belief—or doubt—for that matter? Did we now have the tools to begin to scientifically understand the experiences that give meaning and purpose to so many people and which have played such an enormous role in shaping our history and culture?

As is so often the case with science, I was not pondering these questions alone. Although neurologists were hesitant to study near-death or other spiritual experiences, unfortunately that was not the case with specialists in other fields. I watched with both wry amusement and professional concern as cardiologists, radiologists, and cancer specialists speculated wildly about brain activity during near-death experiences. I was dismayed when their misuse of science led to what I knew were misunderstandings and myths: people returning from brain death miraculously intact, or near-death experiences that *proved* God exists and we are all headed for an afterlife.

It was hardly coincidental that all this heated speculation about near-death experiences took place against the backdrop of our ongoing culture wars. There is a widening schism between people who think God is an anachronism and regard all spiritual experience a dangerous delusion and those who consider religion the core of their lives.

Watching this painful national drama play out from my position as a health-care executive, researcher, clinician, and professor at the University of Kentucky, I was determined that someone strongly based in neuroscience, who knows how the brain works, should try to explain the nature of spiritual experience, not explain it away. The question Joe's experience had posed, along with all the other accounts of near-death experience I had collected over the years, it seemed to me, might be the key to unlocking at least some of the brain processes behind this tremendously important part of our humanity.

Of course I was not the first person to try to locate the divine or the soul in the brain. That quest probably goes back to the dawn of human evolution. There is convincing archaeological evidence in a wide spectrum of prehistoric societies of trepanning, crude brain surgery to open the skull. We are fairly sure this had to do with early man's exploration of the relationship between the spirit world and the brain.

The Greek father of modern medicine, Hippocrates, was an early proponent of what has proved to be a very popular idea: locating the spiritual in the cerebral cortex, the large, convoluted, lobed, walnut-shaped part of our brains that sits under our skulls. The cortex is the brain's most highly developed and recently evolved structure, responsible for language production, memory, reason, problem solving, and what neuropsychologists call "executive decision-making": the qualities that most clearly distinguish us from other animals.

Descartes disagreed with Hippocrates. He identified the pineal region as "the seat of the soul," the transducer that brought the divine, immaterial aspects of the universe into the material realm. It's interesting that modern neurology calls the pineal our "organ of darkness."

Billions of years ago in amphibians and reptiles, it was a patch of photosensitive cells on the top of the head, a "third eye" that in the course of evolution dropped down into the skull in birds and mammals and is now responsible in human beings for the production of melatonin.

Descartes notwithstanding, the bias toward the cerebral cortex as "the seat of the soul" continued through the nineteenth century. Neuroscience giant Paul Broca, the father of modern localization, was entirely consumed with the cerebral cortex, and that emphasis continues in the field today. Understandably so. The cortex puts so many elements of human experience together.

I shared this bias as I began in earnest to look for a scientific approach to understand Joe's ICU experience, but my particular academic orientation gave me a very different perspective. You might say I tended to think outside the cortex. My emphasis had been on the brain regions and processes that control the autonomic nervous centers: the heart and lungs, which are tightly linked to emotions and our survival reflexes. These are in the more primitive parts of the brain, the brainstem and limbic system. As a neurologist, I was coming from the bottom up rather than the top down. This, by the way, is an approach to human nature pioneered by William James—the towering nineteenth-century American philosopher and psychologist—that had largely fallen by the wayside in the ebb and flow of scientific ideas and emphasis. We shall hear more about James, author of *The Varieties of Religious Experience*, later; for the moment, suffice it to say that my orientation toward the brainstem, identified in the ancient Chinese philosophy of Taoism as "the Mouth of God," was fortuitous. Chance, as they say, favors the prepared mind.

Early one sunny summer Sunday morning in 2003, I was sitting in front of an open window, perusing *Life After Life*, a seminal work on near-death experience published in 1975 by psychiatrist Raymond Moody. The book's case studies focused on people who felt that their near-death experience had given them a glimpse of heaven, God, or some other manifestation of a spiritual reality that underlies our physical lives. I was surprised and disappointed at what few medical facts Moody provided, although this disappointment was tempered by the pleasure of reading his nonjudgmental retelling of these extraordinary tales. I analyzed each narrative with the eye of a clinical neurophysiologist, hunting for clues as to how the brain worked during these experiences. Something extraordinary happened to me as I read about the case of Mrs. Martin, who described her experience to Moody during an unexpected cardiac arrest:

"I heard the radiologist who was working on me go over to the telephone, and I heard very clearly as he dialed it. I heard him say, 'Dr. James, I've killed your patient Mrs. Martin.' And I knew I wasn't dead. *I tried to move or to let them know, but I couldn't.*" (Italics mine.) As I read this passage, I was struck by the way that Mrs. Martin appeared to be awake and fully conscious of the world around her, just like Joe, although her muscles seemed to be paralyzed. There was no reason to believe she had received paralytic drugs. So I asked myself: what natural physiological process could have caused her precipitous, total, yet temporary paralysis?

Suddenly, it struck me: we all experience paralysis many times each night, during that period of our slumber when our eyes move rapidly beneath our lids, as if watching events before us. It is called the rapid eye movement stage of sleep, familiar to everyone who studies the brain. We call it the REM state of consciousness.

My breathing quickened. I became light-headed. A cascade of tremendously exciting associations immediately followed that caused my heart to pound. I quickly saw many of the hitherto-mysterious and key elements of near-death experiences sliding, like pieces of a puzzle, smoothly into place.

I knew that not only did the REM state explain Mrs. Martin's temporary paralysis, it also pointed directly to the light that beckons toward eternity or expresses the divine and is one of the best-known features of near-death experiences. "Going to the light" has become a cultural cliché and the subject of both parody and reverence.

Light, as I was well aware, appears to us often in the REM state, when our visual system is strongly activated. As we enter REM, electrical waves propagate from deep in our primitive brainstem upward to visual regions in the cerebral cortex, including the occipital lobes in the back of the head, each about the size of a half fist, where visual pathways terminate from the eye.

Could this light also be the light of near-death experiences, coming from the same parts of the brain? With death approaching, what if we were overtaken by REM paralysis, our visual system stimulated to produce light, and the dreaming apparatus in our brains triggered—all while we were consciously awake and in a state of medical crisis? REM consciousness and wakefulness blending into each other as death approached could explain many of the major features of near-death experiences.

I assembled a research team of neuropsychologists to test my hypothesis that connected REM to the near-death experience. We collected one of the largest numbers of research subjects with near-death experiences ever compiled and compared their sleep experiences to the experiences of other people matched by gender and age. What we found intrigued the scientific community and sparked international media attention.

At some point in their lives, those people who have had a near-death experience were also very likely to have experienced some aspect of REM sleep intruding into their everyday waking world. During the transition between being awake and being asleep, many of our research subjects found that they had become momentarily paralyzed or had visual or auditory hallucinations (roughly a quarter of us will experience what we call REM intrusion).

This meant that the brain mechanism that flip-flops us between

REM sleep and wakefulness was different in people who have had a near-death experience. Instead of passing directly between REM sleep and wakefulness, their brains are more likely to fragment and blend those states of consciousness. Paradoxically, when this happens, they find themselves awake *and* in the REM state. They experience light and out-of-body sensations. They are conscious yet unable to move. They participate in stunningly imaginative narratives. *All* these key features of the near-death experience could be traced to REM.

It is well established in neurology that the mechanism with its flip-flop switch is either in the REM-on or REM-off position well over 99 percent of the time, although on rare occasions, and not in everyone, it gets stuck somewhere in between. This is precisely what we saw happening in our research subjects.

For the first time, our research gave us insight into how the brain works during a near-death experience; we had a hypothesis that could be scientifically tested and confirmed. Although I knew our research was provocative, I was stunned at the furor it created in the popular press. It seems we want to know, but we don't trust what anyone else says about our most personal spiritual experiences.

In the following pages, I describe my ongoing research into near-death experience and what I see as related spiritual events: out-of-body experiences, feelings of rapture or nirvana, mystical "oneness," and visions of saints or the dead. I show how activity in the primitive brainstem, working in tandem with the limbic system, the most ancient area of our recently evolved cerebral cortex, leads to a variety of spiritual experiences.

I expect my work to continue to generate controversy. On one hand, the link I have made between REM and the near-death experience upsets those who see such experiences as a revelation of the afterlife or proof of an underlying web of consciousness or the existence of God. For these people, my work puts near-death experiences uncomfortably close to dreams—in other words, experiences that aren't real.

On the other hand, my work also irks some die-hard atheists, because it inextricably links spirituality with what it means to be human and makes it an intergral part of all of us, whether our reasoning brain likes it or not.

Our spiritual experiences have instinctual qualities, originating in the most primitive parts of our brains. They appear intertwined with the brain's limbic structures, which produce feelings and emotions. My research sheds new light on the irrational, primal nature of spiritual experience and religion.

But I do not believe the primal brainstem and limbic system are the be all and end all of the spiritual. Although the basis for our spirituality can be primal, what we do with it is another matter. The future neuroscience of the spiritual may help us distinguish the primal roots of our spiritual impulses from the associations, imagery, and thoughts they induce in "higher" brain regions. We may finally begin to understand how the spiritual has shaped the cerebral cortex.

This book is only a beginning. The field of spirituality and the brain is in its inception. Each of us, on our own, must find spiritual meaning and value. This is one of our greatest burdens, but it is also one of our greatest opportunities. In the end, understanding the neurological foundation of spirituality is necessary for a contemporary understanding of what it means to be human.

PART ONE

THE ARCHITECTURE

I

WHAT IS A SPIRITUAL EXPERIENCE?

FROM FEAR, TO PINBALL, TO FIELDS OF DAISIES

"And his pure brain, which some suppose the soul's frail dwelling-house."

—Shakespeare, Prince Henry in *King John*, Act vi, Scene vii

"All mental processes, even the most complex psychological processes, derive from operations of the brain."

—Eric R. Kandel, Neuroscientist, Nobel Laureate

It happens in the brain. We may never understand everything about spiritual experiences: there are things about them that may not, in a philosophical sense, be knowable. However, our spiritual experiences depend, as Shakespeare sensed and Eric Kandel writes, upon the operations of our brain. Whether we're believers or nonbelievers; whether we believe, as Joe did, that we have *actually* witnessed Christ and Satan battling for our soul or think that it was merely a hallucination; whether we think the brain creates an illusion of God or believe it is a receptacle for something untouchable and absolute, we should be able to agree the brain is the seat of spiritual experience.

Lilah, my sister-in-law, has pointed out to me that what the brain is doing during spiritual experience may not, in the end, really matter—it's what the experience means to us, how it impacts our lives, that's important. This was driven home to her when her father, Jack, had a series of heart attacks. During one of them, he went into cardiac arrest just after he was placed, awake, on the operating table.

"He told me that he left his body and felt himself moving toward a warm, glowing light," said Lilah. "He felt a great sense of calm and absolutely no fear."

It's interesting that as is the case with many people who have this kind of near-death experience, Lilah reported that Jack wanted very much to *remain* in this place of serenity and warmth. In Jack's case, he was drawn back into his body as he saw himself being shocked back into life by medical personnel with paddles.

"Dad was never a religious man in the traditional sense," Lilah told me. After his experience "he lived with a vengeance, doing exactly what he wanted without any fear of dying. He never set foot in a church. He exuded strength and independence." Nonetheless, she said, her father took great comfort in knowing that "dying would be peaceful and he would return to the 'good energy' that he had experienced when his heart had stopped.

"It really doesn't matter what the brain is doing," she said. "Even if what Dad experienced was some sort of illusion, I'm thankful for it. Not only did it help him, it took away my fear that he would suffer at the end."

Although Jack wasn't a religious man, the family agreed that he had a spiritual experience. But what about it, exactly, made it spiritual? The word "spiritual" is bandied about today and means different things to different people. It originates from the Latin word *spiritus*, meaning "breath or life," which makes sense since our breath, an unseen but vital force, animates us. "Spirit" can be, on one hand, a synonym for the living soul; on the other, the ghost of a person who haunts the world after death. It can mean courage, determination, and energy: "That thoroughbred horse has great spirit!" for instance. Or it can signify mood, such as "the spirit of the occasion." It can point to a deep principle, such as "the spirit of the law." Or it can represent the divine with a capital "S": the Holy Spirit.

The term "spiritual" is generally applied to any human essence connecting us to an unseen world that defies scientific measurement but which we nonetheless believe and feel exists, leaving traces here and there. It can refer to anything transcendent or something that deeply moves or transports us and connects us in one way or another to something larger than ourselves. And it is this last transcendent sense of experiencing something beyond us that I will focus on when we talk about spiritual experience.

From the standpoint of a neurologist, understanding the range of spiritual experience and the ways it might manifest in the brain is daunting. Following are some of the questions I asked myself as I collected cases of spiritual experiences over the years:

What triggers spiritual experiences?

What determines their duration?

What makes them more or less intense?

What brings about their end?

What does a person recall or forget about them?

Is there a set of unusual "symptoms" that leads to them?

Do "symptoms" not usually associated with the spiritual—sweating, pain, nausea, or changes in visual or auditory perception—appear in a significant number of cases?

Are they more likely to occur in a certain posture?

These questions look for patterns and overlaps in brain function in seemingly different types of experiences. Let's consider two cases that I think most of us would agree have spiritual elements. Yet they had very different triggers, and the mood of each is remarkably different. They sharply contrast in psychology, neurochemical influences, and how "the spiritual" was perceived.

Fear, Crisis, Then Revelation

When Cliff, a rehabilitation physician in Kentucky, heard from a colleague that I was collecting case studies of spiritual experience, he contacted me and told me his story. He had a patient, a nineteen-year-old male, who was walking home drunk from a college party one Saturday night when he was struck by a hit-and-run driver, suffered a severe brain injury, and was comatose for several weeks. When he regained consciousness, he was partially paralyzed and was referred to Cliff for rehabilitation.

"Although he eventually recovered sufficiently to ambulate with assistance," said Cliff, "his parents were heartbroken and furious that their son had been injured. For some reason, they directed their anger at me over a minor complication in his recovery and sued the hospital, a suit that was eventually withdrawn because, frankly, it was baseless. Although I knew I hadn't done anything wrong, I had to go through several heated meetings with the parents where their lawyers pushed me pretty hard. The hospital's legal team was there as well. My professional reputation was on the line."

As a medical staff leader, I am often called on to step into just these types of crisis situations between physician and patient (or the patient's family). I know how explosive and frightening they can be for everyone involved.

Cliff said that before one of the meetings with the family he was really keyed up. "My heart was racing," he recalled. "I felt light-headed, in a sweat. We waited and waited for the family and their attorney to show up. I felt unbearably tense. Finally, they called to say they couldn't make the meeting. The oddest thing happened. I felt a tremendous release and abruptly entered a fugue state. I felt dissociated, removed from immediate reality. And then, out of the blue, I was consumed by doubt. Did God exist? I struggled with that uncertainty. My doubt led to feelings of deep religiosity—a major connection to something larger than myself, the feeling that I had had a transcendental insight into the nature of things, and a deep conviction about God's central role in the world."

Cliff said he had a "brightening of vision" as though his vision had "intensified." He described his vision at the beginning of the episode as "washed out."

When the experience began, Cliff was standing. People around him realized he was acting abnormally, and they quickly ushered him to a back room, where he sat down. He was sweaty and flushed. His blood pressure was highly elevated, which persisted for some time, and his pupils were enlarged.

Cliff was "in and out" of his fugue state for about forty minutes. This was a singular event: he had never experienced anything like it and it didn't reoccur. Neurological testing, routine after unusual "spells," with or without spiritual content, confirmed that Cliff's brain was healthy.

Cliff's experience seems obviously spiritual on first glance, but what makes it so? It wasn't spiritual because Cliff was thinking about God. I can think about God right now, but I am not experiencing anything like Cliff's crisis of faith and subsequent epiphany. Cliff had thoughts of God with very powerful *feelings*, which led to an important *insight*—a profound inner truth about the nature of God. This was by no means an intellectual process.

What caused Cliff's experience? Why did it take the form it did? Why was it so singular, without precedent or recurrence? Cliff's clear physiological changes offer important clues into what was happening in his brain. An unpleasant legal situation with attorneys present in a hospital conference room may seem like a strange setting for a transcendental experience. You might think that down the hall in the ICU where patients like Joe lie near death is where spiritual experiences should occur. Yet it may not be the *magnitude* of danger that ignites spiritual experiences in the brain—abruptness, buildup, or release of fear may be the key triggers.

Religious readers may at this point respond, *Well, the trigger was that God chose that moment.* Maybe so, but if there is a pattern here, it is a scientist's duty to seek and discern its underlying principle.

Let's take a look at another spiritual experience under conditions that are very different from those that may have triggered what happened to Cliff. Some readers may find its authenticity dubious.

Pinball and Absolute Power

When I was a neurology resident in training at the University of New Mexico, I worked with Dave, an intern rotating with me on the neurology service. As part of his training in internal medicine, Dave was required to learn how to care for neurology patients, and as the senior neurology resident, I was responsible for teaching and supervising him. We sometimes finished our duties early and sat out of the way at the nursing station, drinking coffee and talking about patients, sports, and, of course, the nurses.

One day our conversation took an unexpected turn. I'm not sure how we got on the topic, but I found myself telling this talented young intern that I was on the lookout for cases of spiritual experience. Perhaps sensing my clinical, nonjudgmental approach, Dave said that when he was an undergraduate student at the university he had had what he considered to be a spiritual experience, although I might think it was odd. He had been finishing up his pre-med studies one Friday in May when his friend Will arrived from New Mexico Tech in Socorro, some eighty miles south of Albuquerque.

"Will's classes had ended for the year, and we were both ready to party," Dave rather sheepishly confessed. "We dropped some acid and smoked some grass. Then we went out to a pinball arcade down the street. After a short warm-up, we headed to our favorite machine."

Dave hesitated. I reassured him that important medical facts were often uncovered in the most unusual ways and from the most unlikely sources. Sensing that my interest was genuine, he continued: "The pinball machine was located just below a large speaker suspended from the wall that pounded out tunes. The machine had multiple levels and two sets of flippers. It released more than one ball. Will was at the controls, working the flippers and bumping the machine to keep the heavy silver ball in play. The ball streaked around, hitting targets and racking up points. The machine released a second ball, and then the action really heated up. The Who song 'Won't Get Fooled Again' blasted from the speaker overhead. Then the most amazing thing happened—the two balls began to repeatedly hit the bumpers and flippers in perfect synchronicity with Pete Townshend's power chords."

The nurses were busy about their duties, but one or two looked over curiously as Dave became animated in the telling of his tale. He continued in a lower voice. "Will kept the ball in play in precise syncopation with the song's guitar solo. The amazing synchronicity reached a kind of climax of perfection as Roger Daltrey belted out the line, 'Meet the new boss, same as the old boss.' When I heard those words I had the instant, overwhelming sense of being swept up into the absolute and infinite power controlling the universe—the 'Boss' who is in charge of everything and never changes."

"How long did that experience last?" I asked.

"A few seconds. The ball was rocketing around the table. Will's run abruptly ended when the ball shot between the flippers on cue with the music. We looked at each other, realizing that we had witnessed an extraordinary event. I don't know if Will had the same experience of absolute power and truth that I had. But he had been aware of the same stunning harmony between the elements of the song and the rhythm of the ball."

"What do you think about the experience now?"

"I'm not sure. I know that I touched something special, something that remains profoundly true for me even today."

No one would have suspected that Dave, who appeared so traditional, had such a past. He was professional in dress and manner, and I found him reliable and hardworking. I offered him little feedback. I was focused on taking the story in, gathering information. Later in the day I thought that I should have offered him more appreciation. He had opened up to me about a very private event.

Let's assume the synchronization between ball and music was not real. Dave touched something special through the coincidence of pinball movement, music, and lyrics that was transformed by drugs into something transcendent. Would my colleagues in neurology then (or now) agree that Dave's was an authentic spiritual experience even though it was probably drug-induced and neurochemically enhanced? Do "artificial" triggers or enhancements matter when we evaluate the authenticity of spiritual experience? Or is the transcending experience all that really matters, as my sister-in-law Lilah might say?

A Resurgence of "Subjective" Science

I approach Dave's and Cliff's experiences with a clinician's eye; neither were transported to "another world" or saw archetypal figures the way Joe did. They did not experience Jack's "light." Although the circumstances were different in both cases, their spiritual experiences occurred when they were standing and awake; they both felt something powerful beyond themselves.

The Greeks, especially Hippocrates, introduced the art of clinical observation. As doctors do today, they listened to a patient's symptoms and examined the body for telltale signs of injury or disease. They collected cases, organizing and categorizing them. This formed the basis of Western medicine for two thousand years.

As a method to investigate and understand disease, the art of clinical case studies peaked in the nineteenth century. Neurological pioneers sometimes sound like their fictitious contemporary, Sherlock Holmes, in what were the first modern descriptions of many diseases of the brain that we recognize today. In London, John Hughlings Jackson studied complex epileptic spells and alterations of consciousness, which he called "dreamy states." Jean-Martin Charcot's renowned descriptions of neurological disease, culled from subjects in a huge asylum in Paris, lured a young Sigmund Freud from Vienna. It was Charcot's cases of "hysteria" that eventually led Freud to study the psychological mind and found the field of psychiatry, which is tightly linked to neurology (in the United States today, psychiatrists and neurologists are bound to the same professional organization: the American Board of Psychiatry and Neurology). Paul Broca, another eminent Parisian neurologist, discovered through the detailed study of his patient "Tan" that spoken language was localized to the left side of the brain. Although Broca's patient's true name was Leborgne, he was known as Tan because that was the only word he could say for his name.

Until recently, clinical description has largely withered from its peak. Modern medicine emphasizes disease mechanisms, objective data, and technology, which, until recently, tended to keep patients at arm's length. Today, ironically, it is machines that are spurring a renaissance in the neuroscience of first person experience. Special MRI scans image our working brains. They are based on the principle that activity in any given brain area increases blood flow, which these scans show with remarkable precision.

While these images can be truly dazzling, we must recognize their limitations. Much important brain cell activity is *not reflected in blood flow*. Cell clusters—that are important during a spiritual experience, for example—may be too small to register on an MRI. Also, the MRI may not be able to detect when parts of the brain shut down and are *inhibited* from action, which is one of the fundamental ways our brains work.

What if the neuroscientist is mistaken about what a research subject or patient is thinking, feeling, or dreaming at the precise moment the image was snapped? Mary Helen Immordino-Yang and her colleagues used MRI technology to detect feelings of admiration and compassion. Subjects listened to true stories meant to evoke these feelings while undergoing scans. Activity was shown in characteristic regions. But unless the subjects actually experienced and faithfully reported these emotions to investigators, the MRI findings would be meaningless.

In a similar experiment, investigators Sam Harris, Sameer Sheth, and Mark Cohen at UCLA looked for the brain's circuitry for belief, disbelief, and uncertainty. Subjects responded to questions on mathematics, autobiography, and religious topics. The brain regions activated included the prefrontal, important in self-perception. Interestingly, the default brain state is belief. It takes more brain activity to work out if a statement is false than it does to decide it's true. The same investigative team found the prefrontal region had a stronger signal in Christians than in nonbelievers when both were asked questions about God and Virgin birth.

Of course a neurologist *expects* to find brain regions corresponding to belief. But does the brain process religious belief in the same way as it does nonreligious belief? Is believing that Moses received the Ten Commandments on Mount Sinai any different to the brain than believing that Apple computers are superior to Dell?

While these findings are fascinating, they remain true if and only if the subjects whose heads are in the MRI scanner faithfully report their subjective experience. What is true for admiration, compassion, belief, and disbelief is particularly true for spiritual experience. We must be sure that a person is having a genuine spiritual experience at the exact moment the MRI image is taken if we think that the activity we see on the scan is meaningful. This requires that the spiritual experience be precisely identified and measured.

In those few moments when neuroscientists think and write about spiritual experience, they nearly always begin where William James left off. He remains one of the masters of psychological observation, from the nineteenth century, when that art was at its zenith.

Spiritual or Religious Experience?

James, author of *The Varieties of Religious Experience*, published in 1902, was the oldest son in an extraordinary American family. His father, Henry Sr., was a wealthy and influential nineteenth-century intellectual and theologian. His younger brother, Henry James, was a great American novelist; Ralph Waldo Emerson was his godfather.

William, a physician, never practiced. He spent his career writing and teaching at Harvard. His students included Theodore Roosevelt, George Santayana, Gertrude Stein, and Walter B. Cannon, the scientist who, as we shall see, discovered the physiology of life-threatening crisis crucial to the biology of near-death experiences. James's curiosity and experimental daring led him to investigate subjects—including religion and spirituality—that were often dismissed by the scientists of his day. He influenced modern ideas of consciousness; in fact, it was James who coined the term "stream of consciousness," which had a profound impact on both the emerging science of psychology and on Modernist authors such as James Joyce, Virginia Woolf, and William Faulkner.

Many of James's ideas on science and spirituality are the foundation neuroscientists continue to build on. Particularly important is *The Varieties of Religious Experience.* Michael Trimble, an expert in both neurology and psychiatry, calls it "the most revealing investigation into the psychology of religion ever attempted." The first chapter, titled "Religion and Neurology," shows James's early recognition of the important relationship between the cerebral and the ethereal. Most of James's contemporaries looked askance at this connection—both scientists and spiritualists. James titled the book *Varieties of Religious* rather than *Spiritual Experience* probably, at least in part, to distinguish his subject from the popular nineteenth-century "spiritualist" movement of mediums and clairvoyants, which was rife with charlatans. James did attend séances held by leading mediums; he wanted to see for himself if they opened a "subliminal door" into the reality of an unseen universe, "a wider world of being than that of our everyday consciousness."

Today, when speaking about a personal experience, people often use the word "spiritual" in the way that James used the word "religious." Historically, much of what would now be spiritual experience happened within established religions (the word "spirit," for example, is referred to throughout the Judeo-Christian Bible). But with the ascent of secularism in the late nineteenth and early twentieth centuries, the word "spiritual" acquired meanings outside of organized religion.

In this book, "spiritual" refers to direct personal experience, regardless of social context. Distinguishing the spiritual from the religious can be difficult, but restricting "religious" to a collection of brains in social contexts (such as being in an organization with traditions, customs, doctrines, and creeds) allows us to focus on the spirituality in individual brains.

So what makes a personal experience spiritual? For William James it was "the feelings, acts, and experiences" of individual people who in their solitude understand that they have touched "whatever they may consider the divine." This definition is both inclusive and expansive. It means many things to many people.

But an implicit and essential feature of spiritual experiences is that they are exclusive to individuals and are not shared directly with others. While members of a group may simultaneously reach ecstatic states (in a revival tent, for example, or when dervishes whirl together), each individual's experience remains very much his or her own. These individual experiences contribute to the structure, doctrines, and creeds of religious organizations. In James's words: "Founders of every church owed their power originally to the fact of their direct personal communication with the divine."

This insight was reinforced in the 1960s by the famous psychiatrist A. H. Maslow, who wrote of the spiritual as "peak" experiences. He, too, thought that all religions have at their core "the private, lonely, personal illumination, revelation, or ecstasy of some acutely sensitive prophet or seer. . . . Organized religion can be thought of as an effort to communicate peak-experiences to non-peakers, to teach them." Maslow, by the way, noted that Americans in the early 1960s compartmentalized religion and religious activities and distinguished them from spiritual experiences.

So what makes a personal experience spiritual? James came up with various elastic criteria. He studied case histories and focused on an individual's "original experiences that were the pattern-setters" in his pursuit of a rigorously defined set of features that all spiritual experiences share. He sought cases that could serve as benchmarks, drawing heavily on spiritual experience that had been documented by psychologists of religion who were his contemporaries. He was less interested in premade or "secondhand" spiritual knowledge and beliefs handed down by parents or borrowed from others—what we might think of as the creeds of organized religions.

A Notorious Case

One of the case studies featured in *The Varieties of Religious Experience* was notorious in its day. It illustrated for James the type of spiritual experience that he came to believe was the "root and center" for all such experiences.

John Addington Symonds was a renowned Victorian art historian and Renaissance scholar. In autobiographical writings, he described unusual dreamlike states in his youth "approaching hypnotism in its character":

> When I was reading, and always, I think, when my muscles were at rest, I felt the approach of the mood. Irresistibly it took possession of my mind and will, lasted what seemed an eternity, and disappeared in a series of rapid sensations. . . . The universe became without form and void of content. But self persisted formidable in its vivid keenness, feeling the most poignant doubt. . . . And what then? The apprehension of a coming dissolution, the grim conviction that this state was the last state of the conscious self, the sense that I had followed the last thread of being to the verge of the abyss, and had arrived at the demonstration of eternal Maya or illusion. . . . The return to ordinary conditions of sentient existence began by my first recovering the power of touch. . . . At last I felt myself once more a human being.

James noted that in Symonds's spiritual experience "there is certainly something suggestive of pathology." Symonds contributed to this view: he suffered from real or imagined maladies all his life and wrote (sometimes beautiful) descriptions of migraine headaches before physicians knew what they were.

Symonds was anguished and confused about the origin of his "trances." He wasn't sure which was real: his normal state of mind or the heightened trance states. These "spells," however, did contribute to his ever-present sense that there were unseen worlds behind everyday consciousness.

James considered Symonds's experience important and authentic, but other influential contemporaries had a far more negative view of Symonds's "moods." In the "Cavendish Lecture on Dreamy Mental States," published in the prestigious medical journal the *Lancet*, the highly regarded English neurologist Sir James Crichton-Browne wrote that Symonds's "life has been described as a great spiritual tragedy, and it was so apparently because his highest nerve centers [brain] were in some degree enfeebled or damaged by these dreamy mental states which afflicted him so grievously."

This assessment of Symonds's "mental states" was nastier than it first sounds. Crichton-Browne borrowed the term "dreamy" from his contemporary John Hughlings Jackson to describe epileptic seizures arising from the limbic or emotional system. Limbic epileptic seizures can alter consciousness, and Crichton-Browne considered epilepsy "a blighting, a crippling, a destroying disease" and claimed that it had left "a permanent mark or blemish" on Symonds.

We now know that abnormal electrical activity in the brain's limbic system can produce facets of what Symonds describes, but we have little to go on to prove he actually had epilepsy. The evidence does not point to seizures; nor does it point to brain degeneration. During the period of his "trances," Symonds received several awards for outstanding scholarship at Oxford, including the prestigious Chancellor's English Essay, which launched his career as a scholar of international stature. Crichton-Browne's opinion of Symonds's experiences is one of the first times a neurologist with a modern understanding of the brain grappled with the spiritual, and he does it badly.

The ad hominem attack by Crichton-Browne on Symonds's

"blemish" and the authenticity of his spiritual experiences may have been a Trojan horse. Symonds published highly controversial writings defending homosexuality (one essay was titled "Male Love"). We do not know what was behind Crichton-Browne's opinion of Symonds, but we do know that Symonds's avowed homosexuality scandalized Victorian England and, along with his well-known "trances," left him vulnerable to ridicule and attack.

A Field of Daisies

There is a risk to talking about one's spiritual experiences. It persists even today and is borne out by my research and clinical experience. In many cases, people consider their spiritual experiences as oddities that indicate a brain disorder or a lapse in sanity or at least sound mind. Neurological disease and spiritual experience occasionally do go hand in hand (as we'll see later). Cliff sought a neurological evaluation after his fugue state, and Symonds drew the attention of one of Victorian England's few practicing neurologists. If one's brain is somehow disordered or diseased, then the experience itself is diseased, and no one wants diseased spirituality.

Many people who have had spiritual experiences have come to me seeking a scientific understanding of what has happened to them. Cynthia, a pediatrician, excitedly contacted me after my team and I published our research findings on near-death experience. When she was pregnant with her first child, she had developed a severe urinary tract infection, which landed her in the ER. She went into shock as she was lying on a stretcher, and her blood pressure plummeted. She found herself briefly looking down on her prostrate body until physicians revived her.

Cynthia is extroverted and leads a team of doctors at a premier academic medical center. When I was contacted to assist with a television documentary on near-death experience, I thought she would be a natural. But Cynthia declined. She said that publicly revealing her out-of-body experience might cause people to question her professional competence and integrity.

Not everyone is shy about their near-death experience. The same week that I approached Cynthia for the documentary, a patient of mine told me the following story within minutes of my first meeting her.

Paula said when she was twenty-two her best friend's husband drugged, beat, and raped her, and then left her for dead. She suffered a severe head injury and was comatose for months. Her doctors thought she was going to die, and, later, when that proved less likely, that she would remain comatose. At some point during her improbable recovery Paula had a near-death experience.

"I went to heaven," she told me. "I didn't see a bright light or know how I had been transported there. But I found myself standing at the gates of heaven. My beloved grandfather, Paw-paw, came out of the gates to greet me. He told me that I must go back, that it wasn't my time, and that if I didn't return, my mother would 'just die.' We walked together in a field full of sparkling daisies."

Paula couldn't recall how or when she had returned from heaven. The experience must have been early in her recovery: she recalls that it was before she was asked to follow objects with her eyes and that she was unable to move.

When we met, she lived independently and was engaged. She was seeing me to find out how to treat her residual leg stiffness from her brain injury. ("I walk like an eighty-year-old lady and I'm only twenty-five," she said.) Her neurological examination showed the residue of a savage brain injury and answered my most pressing question: why had she told me the intimate details of her injury and recovery when we first met? The reason for her so quickly taking me into her confidence was clear. Her medical reports indicated that her brain injury had impacted her frontal lobes, which are responsible for social inhibitions and restraint. I examined her more closely for infantile reflexes that are suppressed by the frontal lobe when we mature. These reflexes were unmasked in Paula, giving me further evidence that her frontal lobes had been injured. The mystery was solved.

Patients usually don't divulge their spiritual experiences, even in the confidentiality of the doctor-patient relationship. We hesitate to talk about walking through fields of daisies alongside our dead relatives, no matter how amazing, important, and seemingly incontestable the experience.

But what more fundamental qualities do these sometimes embarrassing experiences have in common? Cynthia, Paula, and many others have their spiritual experience while their brains are most likely impaired, far from operating at their peak. Yet what happens to them leaves lasting spiritual effects. This incongruence is not something a neurologist can overlook.

James had a very clear answer for people who doubted the authenticity of their experiences. "By their fruits ye shall know them, not by their roots," he wrote. In other words, spiritual experiences should be judged by the profundity of their effect on us—not by what causes them.

The Four Qualities

James considered the variety of spiritual experience like Symonds's, diseased or not, so important because it transformed individual lives and could be a catalyst for the foundation of religions. These experiences were "peculiar enough to deserve a special name," and he put them in the special category that he called mystical experience.

James identified four qualities of this kind of experience. First, it is somehow beyond language. When Cliff told me his story, it was quite clear that he struggled to convey in words the emotional intimacy and strength of his feelings, which seemed to encompass awe, joy, bliss, peace, and fear. Symonds wrote that he could not "find words to render [his dreamy states] intelligible." The difficult relationship between language and the mystical gives us an important clue about where in the brain it, and perhaps other spiritual experiences, arise. Simply put, we can downplay the large regions of the brain devoted to language.

The second quality of mystical experience that James identified was the knowledge such an experience imparts: "When he sees all in all, then a man stands above mere understanding." The knowledge is not logical, a mathematical equation. It is an insight that does not require science or reason. Like the ineffable, this knowledge on first glance seems also to arise from the emotional regions of our brain. James wrote that mystical truth "resembles the knowledge given to us in sensations more than that given by conceptual thought." Both Cliff and Dave were convinced they touched an ultimate reality. That "knowledge" stayed with them. This type of experience can carry with it a long-lasting authority that can give new direction and strength to one's life. Bill Wilson, a founder of Alcoholics Anonymous, recognized the transformative potential of spiritual experiences. His program, which he acknowledged was developed directly from James's ideas, used spiritual conversion as a cornerstone of what has proven to be an effective treatment for alcoholism.

The third property of mystical experience that James noted was its brief duration. The intensity of the experience may be partially responsible for its transience. "Although the soul sees this for a certain length of time, it can no more be gazing at it all the time than it could keep gazing at the sun." While the experience itself is evanescent, it leaves an indelible high water mark that is followed by a sustained high tide. Once the experience is over, the feelings associated with it fade, but the memory of them is exceptionally strong. Their brevity and transience helps identify the brain's state during these experiences and distinguishes them from religious customs and practices, intellectual concepts, values, and behaviors.

James's fourth quality of mystical experience was passivity. Although he acknowledged that certain mystical practices such as "fixing the attention" could facilitate the beginnings of the mystical experience, once the mystical state sets in, the person feels as if his or her will is in the complete grasp of a higher power. Certainly, Cliff could not will or control what happened to him. Symonds found "I could not induce [trances] by an act of volition." And Dave was in no position to challenge the presence of Absolute Power manifest in Pete Townshend's chords and the amazing trajectory of a heavy silver ball.

Extraordinary Consciousness

We may be embarrassed about our spiritual experiences. But we remember them. We don't know what they mean; we find it hard even to begin to describe them. And when a researcher such as myself looks at the broad range of spiritual experiences in the human population, a great diversity appears. A person may be engaged in almost anything when the apparently divine comes to visit. Certainly the four properties of mystical experience that James identified are typically present, but some people who confront acute danger and medical crisis have a near-death experience and Jesus pops up on the end of the bed. Sometimes spiritual experiences are provoked by fear without any actual physical trauma or event. As we'll see, the spiritual can be found in seizures, hallucinations, psychotic states, drug experiences.

When some perhaps less traditionally religious people speak of the spiritual, they describe achieving a "higher" or transcendent consciousness. And just about everyone agrees consciousness is what makes a brain a brain. For neurologists, consciousness has a specific meaning, and it is identified with processes that have specific signatures in the brain. Indeed there have been dramatic recent advances in our understanding of how consciousness comes to exist and function.

It now seems reasonable to expect that new knowledge will illuminate the mysterious nature of spiritual experiences to an extent James could not have imagined. James considered Symonds's trance "beyond anything known in ordinary consciousness." We certainly know more about the brain machinery of consciousness than about spiritual experience. Perhaps spiritual experience uses the same brain mechanisms that bring us consciousness—but uses them in a different way.

My research has turned up evidence that the switch that regulates consciousness in our primitive brainstem is different—more apt to get stuck between the REM state and waking—in people who have had near-death experiences. By implication, the way that switch functions might be significant in other types of spiritual experience as well. With that in mind, let's take a look at consciousness and see what that shows us about the nature of spiritual experience, and the way in which it may be hardwired into all of us.

2

THE THREE STATES OF CONSCIOUSNESS

AND WHERE SPIRITUAL AROUSAL HAPPENS

"The whole drift of my education goes to persuade me that the world of our present consciousness is only one out of many worlds of consciousness that exist, and that those other worlds must contain experiences which have a meaning for our life also . . ."

—WILLIAM JAMES, *THE VARIETIES OF RELIGIOUS EXPERIENCE*

Our 100 billion or so brain cells are coincidentally about the same number as the stars in our galaxy, the Milky Way. Instead of the Milky Way's revolving disk, imagine stars forming a tremendous sphere. Whether a star shines or not depends on its connections with upward of one hundred thousand other stars, and the blinking pattern somehow enters "upon some cosmic dance" to bring forth a self-aware consciousness. How do these scintillating neuronal suns within our brain bring about consciousness, and do consciousness and the spiritual intersect?

In discussions of the spiritual the word "consciousness" is bandied about. Consciousness is often written with a capital "C." It becomes a stand-in for God or "God Consciousness" or the Universe (with a capital "U"). It can represent, in Western approaches to Eastern and New Age spiritual paths, "the ground of being," an awareness of the connection of all things that has strong similarities to the mystical. The scientist in me is reserved about this view of consciousness as all-pervasive, all-embracing; the "mind at large." It's nonetheless fascinating that consciousness has these associations, which is, perhaps, an outgrowth of the way the word is used to signal the mind becoming aware of itself and its tantalizing potential to become unbound by time and space.

In clinical neurology, however, consciousness has a clear and concise meaning—it is awareness of oneself and one's surroundings and it engages certain brain systems in a certain order. That said, much about how it functions in the physical brain is mysterious, and of dizzying complexity.

Within consciousness, neurology recognizes three states: wakefulness, REM sleep, and non-REM sleep. The opposite of consciousness is coma. This definition and understanding is, admittedly, narrow and breaks down when pressed in certain directions. Still, I want to show how this approach to consciousness can contribute to our understanding of brain function during spiritual experience. My research has focused on the switch in the brainstem that regulates our three conscious states. And, as I said, it seems that at least some of the spiritual characteristics of near-death experiences may be due to the switch in the brain sticking between the REM and wakeful conscious states.

My research points to the idea that spiritual experience erupts in the borderlands between consciousnesses, unconsciousness, and dreaming—when our consciousness states are not whole but fragmented and blended with one another. Perhaps this notion of *borderlands,* so dubbed by Oliver Sacks, is simply another way of expressing a conscious state that is "higher," "transcendent," or "expanded," a consciousness that touches God or the Universe.

But, for the moment, let's look at how our waking conscious state can often, even to trained medical personnel, be difficult to detect and how that difficulty has led to supernatural explanations of how a patient knew something he or she apparently couldn't have known. The stories of these experiences are amazing no matter how you dissect them.

Jan's Case

Jan was walking through her garage when she brushed up against her husband's revolver, which lay, loaded, on his workbench. She felt the gun's heaviness against her hip and, in slow motion, saw the gun tumble through the air and strike the cement floor. It discharged, its roar bouncing off the walls and cement floor and filling her ears. She felt little pain as the bullet tore into the left side of her abdomen and angled upward to lodge in her right lung. She was rushed to the local emergency room, in shock from massive internal bleeding, and, without delay, airlifted to a regional trauma center. When she arrived there, her blood pressure was barely perceptible.

Jan says she was conscious at this point, aware of the flurry around her but unable to move or respond to the medical team. The medical record shows that she was given a paralytic drug so a breathing tube could be inserted into her windpipe, and anesthetics to put her to sleep. The surgeon rapidly made a large incision to thoroughly explore her abdominal organs for injury, then tugged and probed for hidden wounds through every inch of the twenty-five or so feet of Jan's intestines.

The bullet's trajectory had severed the liver nearly in two; only after great effort was the bleeding stopped. Jan remained in shock. It was urgently necessary to directly inspect her heart for injury: bleeding around the heart can strangle it like an anaconda. Jan's breastbone was hastily sawed in half—the heart lay exposed and isolated, suspended in its sac, beating mechanically. The surgeon cradled it in her hands, squeezed the engorged chambers and was relieved to feel blood flowing freely into the arteries.

From the beginning of these procedures, Jan's physicians had delicately balanced Jan's nearly absent blood pressure against the risk of stronger sedation or pain relief. To give even slightly more anesthesia would have lethally collapsed Jan's blood pressure. Still, the doctors frantically struggling to save her life did not realize that through all the cutting and sawing and handling of her internal organs, *Jan had never been put fully to sleep.* She was awake! Alert to everything the surgeons were doing and saying, yet unable to twitch the smallest muscle, blink her eyes, or murmur the softest of groans.

"The pain was fifteen on a scale of ten when the surgeon's hands were running over my bowels," she told me. "When my heart was being massaged, I became aware of a faint light off to my left. When I noticed it, an incredible sense of love, comfort, and caring washed over me. I then sensed the presence of my deceased mother, telling me it was not yet my time to die and that she would help me. Then I was at peace and mercifully lost consciousness."

Jan's account was verified by her medical record. Her brain's blood supply was so meager during the period of her low blood pressure during surgery that it is remarkable that she remained awake and didn't suffer permanent brain damage. None of Jan's physicians realized during her surgery that she was aware of everything taking place around and inside of her. The senior trauma surgeon was shocked to find out later that Jan was able to recall her thoughts, feelings, and the surgery in minute detail. I have heard many dramatic stories during my years in medicine, including the accounts of two research subjects who had near-death experiences while awake during surgery. But no one recounted the tale with anything like the cool, calm, matter-of-fact tone Jan had. She had a composure and strength you couldn't help but admire. And it seemed to come directly from her spiritual experience.

From a neurological perspective, Jan's spiritual experience occurred as her brain was about to switch off and she was in the *borderland* between wakeful consciousness and unconsciousness—the transition between consciousness and coma. Her brain was not receiving enough nutrients to keep up the activity necessary to keep her conscious.

A Brief History of Consciousness

Philosophers, psychologists, and neuroscientists have together done much to bring human consciousness into focus and reveal many of its not so obvious details. Neuroscience has produced a parade of theories to explain it. There are several properties they more or less agree upon. First, consciousness (at least with a small "c") abides within individuals and is not directly shared. It is stable over time, with memory unifying the past with the present. Its elements dynamically shift to and fro between the subconscious background and conscious foreground. Many contributions come to consciousness from the senses, intertwined with mental processes that include emotions, thought, creativity, memory, and language.

Conscious experience contains elusive subjective aspects, called qualia by some mind-brain scholars. Qualia refer to the subjective aspects of experience, such as the smell of fresh bread, the feel of silk on our fingertips, the blue of the sea. Qualia have been much discussed; they are at the heart of the mind-brain conundrum—how the physical brain can produce the nonphysical qualities of mind. Not everyone agrees about qualia's meaning, which mental states contain them, and what species, from human to cockroach, are capable of having them.

It wasn't until the 1980s that scientific interest in consciousness began to swell, spurred on recently by imaging techniques sufficiently advanced that we can investigate conscious states like compassion or belief as we gaze upon the working brain. We can actually see what parts of the brain are active or inactive as we dream. Now that we have this objective scan, we can enrich it with subjective, first-person experience.

A couple of generations ago James wrote about consciousness: "A genuine glimpse into what it is would be *the* scientific achievement, before which all past achievements would pale. But at present, psychology is in the condition of physics before Galileo."

Technology such as brain imaging has helped bring today's understanding of consciousness far past even the "Galileo era" yearned for by James. Our understanding of consciousness today is more like a Hubble view of the universe than any peek provided by Galileo's telescope. We can peer both at the beginnings of the universe 14 billion years ago and at the workings of the brain, the place our identity emerges.

Consciousness Now

In medical settings, consciousness can seem a routine and then, in the next moment, an urgent question, needing immediate resolution. Patients with disordered consciousness, such as coma, often require decisive action delivered in seconds. In times of medical crisis, consciousness is not a dull philosophical matter.

The neurologist determines a patient's conscious state by the patient's reaction to different stimuli—voice, light touch to deep pain—and by using judgments based on brain anatomy and physiology.

Each of the three conscious states—wakefulness, REM sleep (when we do most of our dreaming), and non-REM sleep—has identifiable brain activity that can be recorded by electroencephalogram

(EEG) and brain scans. Of course, typically you don't need a lot of medical hardware to determine whether someone is awake or asleep. It's in between conscious states and coma where things become tricky.

Seeing consciousness in practical neurological terms has its limits. None of Jan's physicians were aware that she was fully conscious. She couldn't move, and her reactions to the extreme pain were imperceptible. She nonetheless proved herself quite capable of laying down memories of her thoughts and feelings and the events during her surgery in stunning detail. Her recall is proof that Jan had all the necessary features for neurological consciousness, despite a medicated brain that was getting very little blood.

A response from a patient, even a complex one, is not necessarily consciously driven. Some patients in permanent coma will open their eyes, while in others the eyes stay closed. If we deliver a well-placed tap upon the knee to someone who is brain dead (which is when *all* the brain cells have died), it is possible to see the familiar knee-jerk reflex, which comes from the spinal cord, a part of the nervous system that operates below our consciousness. In Jan's case, the paralytic drugs she had been given would have blocked the knee movement, yet, as we know, she was quite conscious. So using physical reactions to determine consciousness can be misleading.

People can appear dead but still be very much alive. The brain can be conscious and quite alert to what's going on around it even when it's seemingly insensible. I have doubts about near-death experiences where people are supposed to be unconscious but still somehow aware of what's going on around them. Imprecision about what consciousness is and isn't has created rampant misunderstandings and myths not only about near-death but about a whole range of spiritual experiences.

A single nerve cell (or neuron) by itself is not conscious. Even a cluster of neurons is not conscious. Activity in a multitude of nerves does not necessarily lead to waking consciousness either, for our brains can be surprisingly active during the conscious state of dreamless sleep and, counterintuitively, in coma as well.

Consciousness is something else—an awareness that is the summation of nerve activity, distributed and processed throughout the brain. Brain activity is necessary but does not necessarily lead to consciousness. We are just learning that in the coma state the brain is remarkably active. What is it doing? We're not sure. But it is not enough activity or the right kind of activity to bring the comatose to consciousness.

Back to the Brain Cell

Over one hundred years ago Charles Sherrington pondered the connection between neurons, reflexes like the knee jerk, and consciousness. Sherrington, who would eventually win a Nobel, was born in London in 1857 and raised in a comfortable middle-class home that treasured scholarship. Early in his career he turned his attention to the spinal cord, which connects our brain with our body below the neck. His discoveries at this primitive level carried through to the brain and consciousness itself.

Sherrington—a short, wiry, shy man with a gentle disposition—had large, powerful ideas. Parallels have been drawn between Albert Einstein, who radically changed how we see the physical universe, and Sherrington, who revolutionized the universe of neuroscience. Their parallel universes crossed in 1924, when Sherrington presented the prestigious Copley Medal to Einstein. We take it for granted now, but in Sherrington's time influential neuroscientists relentlessly argued that specialized regions of brain function were a myth. Instead, the brain cells were physically connected and acted "holistically" as a huge network. In this view, the brain was equipotential throughout, and specialized functions, like say drawing or speaking, were not location-specific.

It was Santiago Ramón y Cajal, Sherrington's inspiration, who found the convincing evidence that the nerve cells, the living units of the brain, were not physically connected but led independent lives. A rebellious and headstrong youth did not portend the brilliance that marked Cajal's career. His father, a physician, realized his wayward son needed a new direction in life and was determined to push Cajal toward medicine. To the surprise of both, this tactic worked. Cajal went on to establish the microscopic foundation for understanding how the brain could sequester clusters of independent nerves to localize a function, perhaps such as spiritual experience. But that left a huge question: how did the nerves communicate if they had no physical connection to produce consciousness?

This physically tiny but conceptually huge gap was bridged largely by Sherrington. He developed the idea that brain cells communicated through junctions or "synapses." Not only that, he pointed out that this communication was *one way*. It went from one brain cell to another, but never reversed and traveled backward. Later it was discovered that nerve cells communicate with one another across these junctions using chemicals transmitted from brain cell to brain cell. These chemicals proved to be instrumental in spiritual experience.

Sherrington discovered that these otherwise sovereign nerve cells behaved reflexively, that is the action of one nerve cell automatically led to the action in another nerve cell. The knee jerk is perhaps the simplest neurologic reflex: its circuitry requires only two nerve cells making a single synaptic connection. The first nerve receives the stimulating knee tap, and this nerve makes its way to connect, or synapse, with a nerve in the spinal cord that then sends an impulse to the muscle that kicks the leg. How vigorously the knee kicks can tell me how the brain is working. Reflexes are often much more intricate, engaging large numbers of neurons in elaborate sequences.

As much as Sherrington changed our thinking about how the brain worked, he wrestled unsuccessfully with a problem, which remains vexing for us today: how is it these individual nerves of the brain, communicating one way across a gap and acting reflexively, transform to that "energy which is mind"?

Sherrington described the brain upon awaking:

> The brain is waking and with it the mind is returning. It is as if the Milky Way entered upon some cosmic dance. Swiftly the head-mass becomes an enchanted loom where millions of flashing shuttles weave a dissolving pattern, always a meaningful pattern though never an abiding one; a shifting harmony of subpatterns.

Sherrington, who lived to ninety-four, was at least partially reluctant to reduce mental activity, let alone spiritual activity, to material explanations of the brain because, he wrote, "our mental experience is not open to observation through a sense organ." The brain cannot grasp itself and be conscious of itself as a brain; instead it is conscious of something else—thoughts and feelings, for example. The brain and mind lived dual lives for Sherrington.

Sherrington might have changed his mind today, now that we have MRI scans, which can become an extended sense organ for consciousness, detecting admiration, compassion, belief, and disbelief. We are closing the gap between mind and brain; it is no longer the yawning abyss it was in Sherrington's time. Futurists predict that soon microscopic machines will move through our bodies, repairing cells and organs. Charles Lieber at Harvard and others are using nanotechnology in a radical new way that allows matter to interact with the brain's energy. With wires only a few nanometers wide (the width of a human hair is about one hundred thousand nanometers), Lieber and his colleagues have managed to detect signals from individual nerves on a circuit board that create a grid of neural reflexes using the tiniest of wires. This is close to mimicking the natural synapses that connect nerves, and perhaps one day these circuits could become part of a feeling, movement, thought, or emotion. Such tools raise many complex issues and new ethical problems as machines become part of our brains, part of our selves.

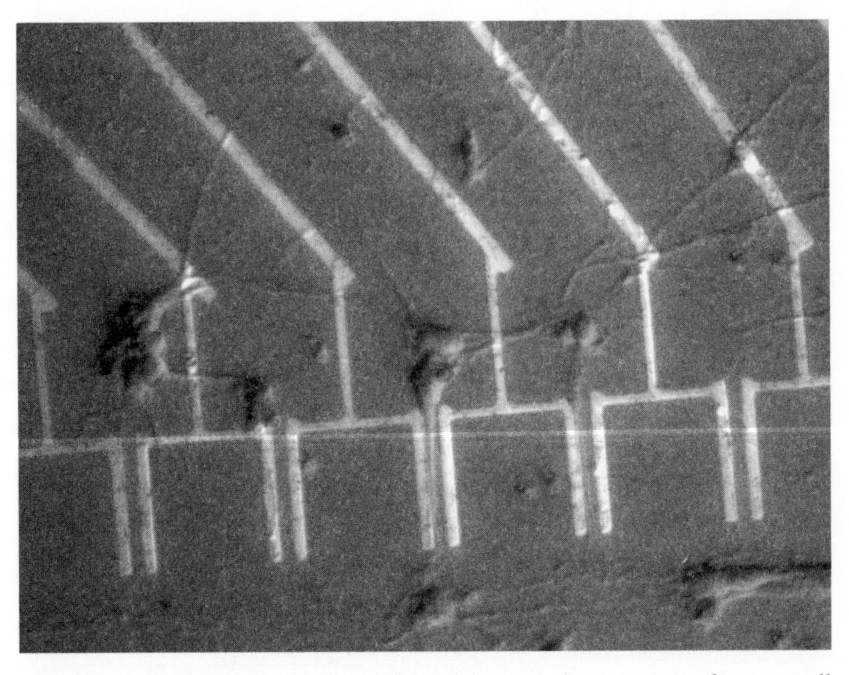

Figure 1: Cultured neurons resting on a grid connected to nanowires far too small to be seen at this magnification. These arrays may permit sending and receiving signals from nerves that closely imitate natural nerve communication.

Aroused to the Spiritual

We must be conscious to have a spiritual *experience*, which may seem obvious but has a hidden significance. In order to be conscious, we must be aroused.

Although the word "arousal" commonly refers to being sexually excited, neuroscientists use the word more loosely to mean that we are readied to receive the internal world of our body and brain and the external world around us, which also regulates the conscious states of sleep and wakefulness.

Arousal is carried out by fundamentally distinct physiological processes in clearly identifiable brain regions.

The nerve centers arousing the brain to respond to itself and the environment, switching the brain to sleep or wakefulness, are in the structure called the brainstem, the most primitive part of our brain, located at the base of the skull. In 1949, Giuseppe Moruzzi and Horace Magoun, neuroscientists working at Northwestern University in Chicago, found that by directly stimulating this tiny area in animals they could activate the much larger cerebral cortex. It is in our cortex that our human attributes arise. Moruzzi and Magoun gave us the first glimpse of how the brainstem orchestrates consciousness. It's just as telling that not only can we stimulate arousal, but when regions within the brainstem only millimeters in size are destroyed, a deep, permanent coma ensues.

The brainstem not only arouses us, it also regulates breathing, heartbeat, "fight or flight," and vegetative functions. The arousal system contains the switches that shift our consciousness between its three states. The body's nervous system connects with the arousal system through the spinal cord that extends upward as a great stalk, entering the skull to merge with the brainstem. This same system automatically controls our heart and lungs.

The brainstem's compactness belies its highly complex anatomy and physiology. It began developing early on in evolution, about 300 million years ago, and it has changed very little from species to species since that time. From rat to human, the brainstems of *all* mammals are curiously similar. The reason for this is that the brainstem's function is so critical—to get it wrong in even the slightest way is usually incompatible with life itself. To Sherrington, ancient regions

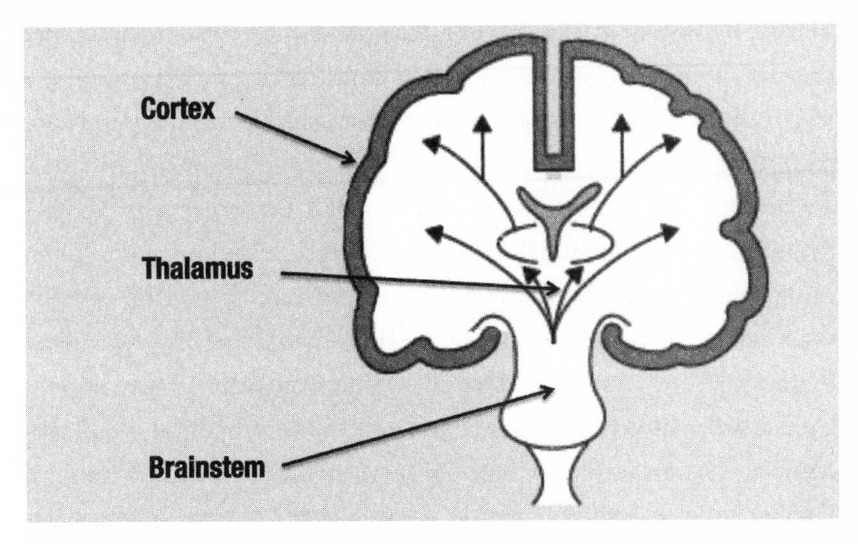

Figure 2: The anatomy of consciousness. During wakefulness and REM (dreaming) sleep, the brainstem arousal system activates the thalamus and cerebral cortex that provides our uniquely human qualities.

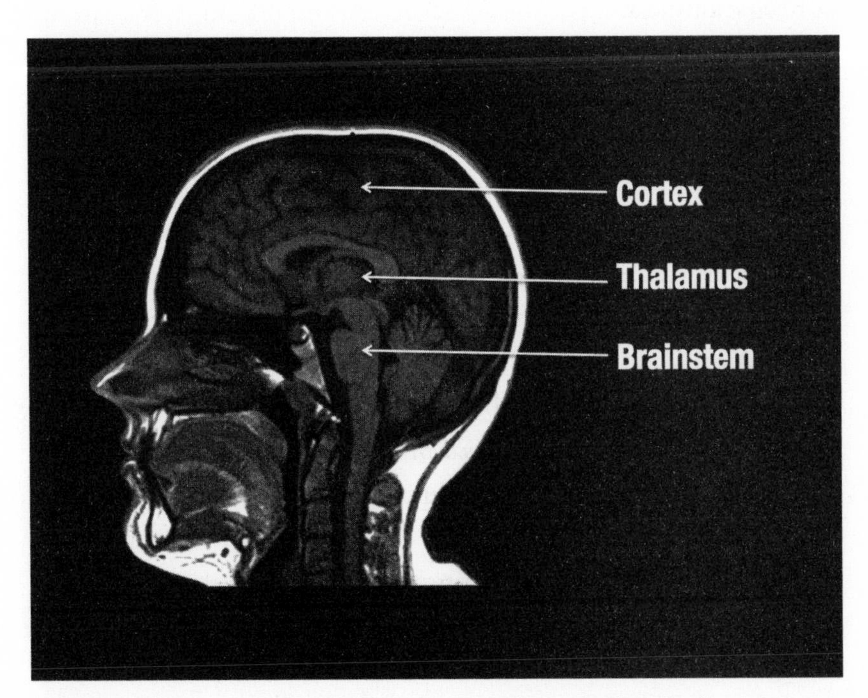

Figure 3: MRI scan showing the brainstem, thalamus, and cerebral cortex

like the brainstem were a "part of our brain which still continues that of man's animal ancestral and related stock of long ago."

Through nerve pathways projecting upward from the brainstem, the arousal system activates many brain regions, using a tightly balanced interplay of the chemicals that span the synaptic junctions between neurons: adrenaline, acetylcholine, serotonin, and dopamine. Human consciousness resides in the cerebral cortex and the thalamus located directly beneath. These two areas are the largest brain structures, with the greatest number of neurons. The cerebral cortex, often called simply cortex or cerebrum, is the most easily recognized portion of the human brain—a massive crown, resembling a convoluted walnutlike mantle above the other brain regions. Its roughly spherical shape can be divided into two great halves, the left and right hemispheres. These hemispheres are not equal in their tasks; they are highly specialized, contributing very different attributes of our consciousness.

Just as a painting's canvas is necessary but is not the painting itself, being aroused and alert is necessary but not sufficient for consciousness. The arousal system readies the brain, much as the canvas receives the artist's brush. The arousal system prepares the thalamus and cortex, which in turn paint the content of our conscious experience.

The cortical hemispheres wrap around the thalamus beneath. The interplay between cortex and thalamus is fundamental to many cerebral cortical functions. All sensory information, except smell, must be relayed by the thalamus before making its way to the cortex to be fully recognized. Bodily sensations—vision, hearing, pain, touch, and limb position—are processed first by the thalamus before being refined by the cortex.

The neurological basis of consciousness is well established. The brainstem awakens the thalamus and cortex above, which in turn bring us the human experience—the mundane and the transcendent. Our brainstem is easily overshadowed by the grander thalamus and cortex; and everyone who has tried to understand the brain in spiritual experience has ignored the brainstem's vital role in regulating consciousness. No more.

The Gateway

Not everything from the external or even internal world makes its way into our consciousness. The brain uses gatekeepers. Think of your right foot. Immediately your right foot moves into your attention's foreground. Although your foot was sending sensations to your thalamus, keeping it abreast of what was happening in your right foot, before you read that last sentence, your attention was focused elsewhere, processing the language of the first sentence of this paragraph (without your attention wandering, I hope). When you ran across the suggestion to think of your right foot, your brain *opened* sensations from that part of your body to reach your consciousness. You were suddenly aware of a new set of information—your foot's position and location, what your big toe was feeling. But all these sensations were present immediately before they entered your consciousness.

For much of the time, it is the thalamus that has the entry key to the brain. The thalamus acts as the principal sensory relay, determining how and what information gets from the foot to the cerebral cortex for conscious processing. When your right foot entered your consciousness, your thalamic gateway opened.

This process of entering consciousness can also be thought of as a spotlight, selecting and retrieving information. Sometimes this spotlight is overactive. Anxiety about an illness can drive the spotlight to shine on many normal sensations that are otherwise disregarded. If a pinched nerve in the back leads to intermittent pain in the right leg, it is natural to start paying attention to the foot, wondering if a feeling here or there means the nerve injury is getting worse. Hypochondriacs are bedeviled by the spotlight: it shines into every nook and cranny of their bodies.

There is no solitary nerve or single nerve cluster in the cerebral cortex that serves as the fountainhead of consciousness. There is no "consciousness center." Most neuroscientists think that human consciousness is a *process* that comes alive by the thalamus and cortex interacting with each other. We know that the thalamus has strong reciprocal connections with the cerebral cortex, forming circuits critical to arousing and regulating consciousness. Damage to this region from strokes or other injury can produce unresponsiveness to the internal and external worlds, even irreversible coma. A permanently closed gate.

What these gates keep out of consciousness may be more important than what they let in during spiritual experience. The gates might be crucial to spiritual experience, but maybe not in the way we at first think.

A Dark Energy

As is probably clear by now, the brain keeps most of its activity under wraps, hidden from consciousness. This is a good thing. No one wants to have to remember to take each breath, quicken the heart rate with excitement or slow it to rest, blink to protect the eyes, coordinate every gland in the gut during a meal, or keep the body's temperature at 98.6 degrees. I want my consciousness free to explore other realms.

Consciousness is not only blind to most brain activity, consciousness is even blind to its own brain process. So consciousness will never fully understand itself, no matter how much we can peek inside the brain.

This does not trouble me. I think it is useful to acknowledge our inherent limits regardless of how much we feel we understand. Most of Einstein's universe and all of any other ones out there will be forever beyond the reach of telescopes and deep space probes; still, awesome splendor rewards what efforts we do make to plumb their depths. We enjoy an expansive view of our elegant universe, although we can sense only a small segment of what surrounds us. Less than 5 percent of the universe is detectable by our instruments; the remainder exists as dark matter or energy. Likewise, some of our consciousness will probably forever be a sort of dark energy to us.

There is much we still don't understand about the way that the cortex brings content to consciousness. It processes an incomprehensible amount of information, arising primarily from the external world, as well as from the *milieu intérieur* of the body and the brain itself. This ocean of information is distilled down to the trickle essential to survival by a process that takes information packets and distributes them throughout many brain regions, where they are simultaneously processed.

This is called parallel distributed processing—multitasking on a massive scale, which seems particularly well suited to our brain with its billions of neurons, many single neurons synapsed upward to thousands of others, conveying excitatory and inhibitory messages. The parallel distributed process allows the same information to interact with many other information units and ultimately permits our brains to assemble information into something that becomes meaningful.

Two Things We Know About Consciousness

Eventually this simultaneous processing must become sequential, since consciousness has a distinct sequential flow, with events strung out along the timeline of our lives. The parallel processing that generates content must be integrated into a sequence for consciousness to happen. This integration takes place in the thalamus, or the connections between the thalamus and the cortex.

The thalamus and cortex act together in a very peculiar manner. They electrically resonate about forty times a second. Why they create this intrinsic rhythm is not clear but might be important in synchronizing perceptions and thinking during wakefulness and dreaming. It is an internal timer of sorts. Possibly this timer helps the brain bind information from distant domains into the whole perceptions that we experience. If it is a rhythm that binds our consciousness, then we know its tempo.

Consciousness is a dynamic process, constantly changing, using nerves distributed throughout the thalamus and cortex. But some neuroscientists think consciousness can be reduced to a lower level in the brain, with some of its aspects perhaps happening in the more primitive brainstem. Christof Koch, a neuroscientist at Caltech, has argued that tiny insect brains, such as cockroaches', "probably [have] what consciousness requires"—neural processes that are "massively parallel and have feedback."

Locating consciousness in the brainstem and attributing it to cockroaches tells me how hard it is to precisely know what we are talking about when we say "consciousness." Plants can "perceive" the sun and track its movement across the heavens, yet I don't think that reducing consciousness to this level serves our purposes in understanding the brain and spiritual experience.

Although parallel process may be necessary for consciousness, with all due respect to cockroaches, this process alone is not sufficient. The Internet has a massive distributed parallel processing capacity, yet we do not consider it conscious (except in science fiction). And other brain structures (such as the cerebellum, which controls balance) are capable of extremely complicated parallel processing and yet have little bearing on consciousness. Regardless of the process that is necessary for human consciousness, nearly all neuroscientists now believe that the thalamus and cortex are the parts of our anatomy where our consciousness attains its unique human content. If we lose these parts of our brain, we lose consciousness that any of us would recognize as human. It seems natural to look for spiritual experience there.

When You Lose Your Brainstem

Maybe so, but the brainstem acts as a primal wellspring of human consciousness. Its pathways carry nerve processes upward from the feet, the gut, the heart, the whole body, to activate the thalamus and cerebral cortex. No brainstem, no consciousness. You may not be dead, but if not, you will be in an irreversible coma.

At least until recently. New technology now offers hope regarding the irreversibility of this state. Dr. Nicholas Schiff and his colleagues at Weill Cornell Medical College in New York cared for a thirty-eight-year-old man who was only minimally conscious after his brainstem arousal system was injured in a physical assault. For six years he was nonverbal and unable to communicate. But when they placed a stimulation electrode precisely in the part of the thalamus connecting to the arousal system, his thalamus and cortex were awakened so to speak, and he became verbal and meaningfully interacted with others around him.

How much brain do we need for consciousness? We could lose half our thalamus and cortex, either the left or right half, and still retain the essential qualities of consciousness. The parts of the cortex and thalamus that consciousness requires are a mystery that brain scans have recently addressed.

If the brainstem of an injured patient remains healthy and can control heartbeat and breathing, yet the person has lost all reactions that signify that the thalamus and cortex are conscious, the brain may be catastrophically injured, in a vegetative state. If the vegetative state persists for twelve months, the prognosis for meaningful brain recovery is beyond dismal, and the coma is permanent. Theresa

Schiavo had this kind of brain injury and became a figure of strident, emotionally wrenching public controversy.

Schiavo's parents tragically denied that their daughter was in a permanent vegetative state. It's easy for families to think that a loved one with a serious brain injury has meaningful responses, even when overwhelming medical evidence tells us otherwise. For all the damaging behavior that followed the Schiavos' tragedy, it was Senator Bill Frist, a heart surgeon, who, from a neurologist's vantage point, made the most egregious error. On the floor of the U.S. Senate, he said that Schiavo was not in a vegetative state. Not only was he wrong and speaking beyond his professional qualifications, but his opinion was formed after he only cursorily reviewed the neurological evidence. In a vegetative state all human qualities of conscious awareness are lost, even though the brainstem's arousal system may still be operating, even though the patient's eyes may spontaneously open. The brainstem alternates between sleep and wakefulness, although it is not wakefulness or sleep that is easily recognized. This state must be carefully distinguished from the minimally conscious state where the injured brain retains at least a slight ability to interact with the outside world. This interaction can include reactions to pain and touch, fixing one's gaze on an object, and responding to questions or commands by word or gesture.

We come to grips with what parts of the brain are required to make us human by understanding the difference between the vegetative and minimally conscious states. Unfortunately, it is still often hard to distinguish them. Bedside reactions are not sufficient, even for seasoned clinicians. This makes things particularly difficult for relatives who believe they see consciousness in reflexive responses like the knee jerk.

MRI scans in both the vegetative and minimally conscious states show reactive brain activity. Recognizing one's name may seem to distinguish the two states, but this test is by no means foolproof, because the thalamus and cortex are not completely dead in the vegetative state and do occasionally react to the outside world. Still, these isolated islands of thalamus and cortex activity are disconnected from other brain functions. This, surely, is *not* consciousness.

Exactly how much cortex and thalamus we need to be human is the frontier of investigations into consciousness. Again, the way forward is using techniques like MRI and PET to help understand what a person is experiencing in certain states of consciousness.

Jan had her spiritual experience in the borderland between consciousness and unconsciousness. This is an important clue to how her brain was working. We lose consciousness in one of two ways. The brainstem arousal system may be injured, as in Dr. Schiff's case, and the arousal system left unable to arouse a working thalamus and cortex. Or, as when the surgeon massaged Jan's heart, the thalamus and cortex may shut down so that consciousness cannot be sustained even when the brainstem arousal is in overdrive.

When one's brain is starved for blood, like Jan's during surgery, a person enters borderlands as wakeful consciousness merges into unconsciousness and sometimes consciousness reemerges again.

A Hidden Relationship

Now that we know a bit about what neurological consciousness is, we can begin to ask if spiritual experience is a unique conscious state, a signature pattern of brain activity that a neurologist would be able to recognize, such as wakefulness or REM sleep. Or if spiritual experience is the expression of a conscious state, like emotions, the ability to speak, the awareness of our body.

If spiritual experience is an expression within consciousness alone, then we should speak of spiritual "awareness" rather than "consciousness" to be correct as far as the brain is concerned. On the few occasions in the history of neuroscience when we have considered spiritual experience, it has been viewed as the stuff of waking consciousness, the product of thalamus and cortex. No one after James has followed through with his idea that spiritual experience, particularly mystical experience, could be a conscious state in its own right. But what if spiritual experience shared the brain mechanisms of waking and dreaming consciousness? What if there was a previously hidden relationship, one between the spiritual and consciousness that James could never have envisioned, a relationship overlooked by neuroscientists preoccupied with the "higher" cerebral cortex? Maybe spiritual experience erupts in the borderlands between consciousnesses, unconsciousness, and dreaming—when our consciousness states are not whole but fragmented and blended: a hybrid.

The borderlands of spiritual experience affect a very special expression of consciousness, the sense of our individual self—the first-person perspective of the "me," which is, except in rare instances, where most of us live.

The self has two special relationships with spiritual experience. First, memories and feelings of a spiritual experience often become among the most important parts of our self-identity. Second, one of the cardinal features of spiritual experience is the loss of self, often into what some people have called an "expanded" state of consciousness.

We need to look at the self from a neurological vantage point, how the self is put together and disassembled. The self can be seemingly lost and then reconstituted in various types of spiritual experience. Neurology can coolly discount the "realer than real" of *any* experience. And what makes it easy is the brain's seemingly limitless ability to create convincing illusions and confabulations that appear to be fact. Such as the notion that you are you.

3

THE FRAGMENTED SELF

HOW WE BECOME OUR OWN FALSE WITNESS

"Know thyself."

—Inscription on the Temple for the Oracle at Delphi

"The most common lie is that with which one lies to oneself."

—Friedrich Nietzsche

An old adage in medicine declares that you can gauge how much we know about a topic by seeing how much is written about it—the more pages produced, the less we understand. The self is a case in point. Many volumes have been written about the self, and many more are waiting in the wings, yet, like consciousness, it is mysterious and elusive, hotly debated and now awesomely arcane.

Transcending the self or "losing" the self has long been thought to be a prerequisite for the mystical experience of oneness. During a near-death experience, what happens to the self is, in many cases, revealing about what's going on in the brain. This chapter does not offer a comprehensive understanding of self and spiritual experience. That is simply not possible. Instead, I look at what we know about the neurological self and how I think that might relate to spiritual experience.

Bat Consciousness

In *The Principles of Psychology*, James first drew attention to the importance of self for understanding behavior and formulated the foundations for its modern study. "Ever since Hume's time, it has been justly regarded as the most puzzling puzzle with which psychology has to deal," he wrote.

In our day-to-day lives, we have an overwhelming sense that we exist as individuals, separate from things around us. It is the sense that it is "me" that is having experiences, such as the experience of reading this book. It is "me" that owns what I experience and is the agent of my actions. The "me" is overwhelming, except, that is, in rare moments of mystical union or deep meditation or certain forms of psychosis. I cannot make it go away and remain conscious. Consciousness and self are almost always intertwined and indistinguishable. But to a neurologist, attuned to brain function, they are different. I look first to the brainstem for consciousness, but I do not search the brainstem when I search for self.

There are ways to separate self and consciousness, including meditation and some drugs. The aim of many types of meditation is to lose the self and in doing so realize the greater Self, the self with a capital "S," which is synonymous with God or capital "C" Consciousness. The Buddhists say that by losing the self we realize our true nature—who we really are.

Under normal circumstances, however, I cannot imagine doing

away with myself any more than I can imagine what it is like to have the consciousness of a bat. When I look at a bat, I can tell if the bat is awake, flying around, or sleeping upside down. If a medical illness should befall the bat, then I might do a neurological examination on it and determine if the bat is conscious or unconscious. At any rate, I can, within the parameters of scientific reason, attribute neurological consciousness to the bat. But I do not directly know the bat's conscious experience. I really don't know what it's like to be a bat, and I can't tell with absolute certainty if there is a bat self or not, a "me" within the bat.

I doubt the bat experiences a "me" in the way I understand "me"-ness. Much of the bat's brain must be devoted to sound so it can "visualize" its nighttime prey in flight. It lives in a very different sensory world from my own, so much so that I cannot even begin to imagine what it is like to have bat experiences. A bat "self" is unlikely and unnecessary to be a full and complete bat in the sense that I understand a bat. Only a few mammals with highly evolved brains can recognize themselves as individuals, but even that is not sufficient for a sense of self. But it is ultimately impossible for us to know if the bat is aware of a separate self unless the bat can directly communicate with us.

We know that the boundaries of self can be distorted or completely lost during mystical experience, an experience similar, if not identical to, the dissolution of the small self in both Eastern and Western meditative traditions. James said the mystical involved overcoming "all the usual barriers [including bodily] between the individual and the Absolute." But dissolving the usual boundaries of the body does not always result in the mystical—or in any kind of spiritual epiphany, for that matter. Cynthia did not have a mystical experience as her consciousness floated around the ER.

Still, the self has boundaries that are often crossed in the borderlands of spiritual experience. It is strange, even paradoxical, that the loss of self becomes one of the most powerful experiences that we can own, usually becoming deeply embedded in our lifelong self-identity. There is no doubt afterward that the "me" of the experience owns the feelings, sensations, knowledge, and aftermath of the types of spiritual experience that involve loss of self! So powerful is this ownership that it often becomes integrated into our enduring religious beliefs and deepest aspirations.

We will examine how it is in the brain that the self breaks down or is conversely reinforced and apply these ideas to different types of spiritual experience. But first we need to know how the brain puts our experience of self together. Some neurological parts of self that are well understood have direct impact on spiritual experience (out-of-body, for example). Our understanding of other components of the self is more speculative. Regardless, we can go no further until we first know as best we can how the self is put together and, sometimes, falls apart.

A Sense of Self

Humans and apes have long been known to see themselves as individuals. Recently it has become clear that other animals with both large brains and well-developed social behavior can know themselves. Dolphins and Asian elephants are able to recognize their own image in a mirror. In addition, wild dolphins identify themselves and other dolphins through unique vocalization labels, just as humans know when their name is uttered by a stranger.

Recognizing the physical self, however, may not be all that important to forming a sense of self. It may be unnecessary for a sense that I am "me." Until relatively recently in human history, the only mirror available to us was our fractured reflection in a pool of water. Many people lived their lives with that blurry picture of their own face—and yet they must have had strong self-identity.

A self that recognizes itself as a distinct individual happens early in a child's development. Self is "a person's essential being," as *The New Oxford Dictionary of English* (1998) puts it, that distinguishes him or her from others. It is a "particular nature or personality"—again from the same dictionary. We will not consider the self in this broad sense: the way we can regard ourselves as unique individuals. Nor will we speak of self-consciousness in cultural or psychological terms or as the awkwardness that strikes with social embarrassment. Instead, we will focus on the neurological facets of the self.

There is an implicit and explicit self. Implicit qualities are those that are parts of who we are but that lie outside consciousness. My DNA defines me, but it is below my consciousness unless I map my genome. Even then I only know a representation of my genetic molecules, not the molecules themselves that I carry around during my "me" experience.

Explicit self is what rises to our consciousness, like your right foot when I draw your attention to it. The border between explicit and implicit self is not fixed: it is ever changing. Loss of self in spiritual experience can be thought of as a severe contraction of explicit self and an expansion of implicit self.

The Everyday Me

What are the "me" typing this manuscript and the "you" reading it? Normally my fingers are invisible to my consciousness as I concentrate on what I want to say. When I do look down and see my fingers tapping away on the keyboard, I know my own hands. I recall them as mine. I can see that they are connected to my arms, which, in turn, are connected to my trunk. The sensations are fused: the fingertips on the keyboard, attached to arms, match my sense of where my hands are in space.

My face is reflected in a corner of the computer monitor. That face is also me. It is the same face I saw this morning in the bathroom mirror when I cut my chin shaving. I can feel the cut. It is also the same face I see when I look at my high school portrait, sitting on the bookshelf. This photo has the same features as my reflection, the same mole near my right sideburn. Today, there are more wrinkles, of course, and grayer hair. I recall driving to the studio in the family car and sitting self-consciously for the portrait.

A hungry feeling comes to me, reminding me of lunch, and I think about preparing the soup I made yesterday. As I get up to go to the kitchen, my daughter calls me on my cell phone. The voice with which I speak to her has been mine for as long as I can remember.

Thanks to memory, I am conserved through time. Memory is the key to the coherence between the external world and what we will be referring to as our "internal milieu." My memory is possible because of stable changes made in Sherrington's synapses between neurons. Those patterns keep me from falling apart after I recover from anesthesia, when my brain and memory have stopped working. If I am to re-emerge from a deep anesthesia, my characteristic nerve discharge patterns must be reestablished after having been obliterated when nerve activity was suppressed. Autobiographical memory includes sensual perceptions, emotions, ideas, and events. Synapses in the left temporal lobes are particularly important for this type of memory.

So important are these synapses that long-term changes in their structure underlie the permanence of memory, the glue that helps hold the self together. When we step back and think about synapses throughout the brain, we see that these channels allowing communication between brain cells are critical to who our brains make us.

As important as memory is to consolidating self, losing great chunks of memory does not mean that we lose ourselves completely and become zombies. A sense of agency and the "me" of experience can persist. Drs. Daniel Tranel and Bradley Hyman at the University

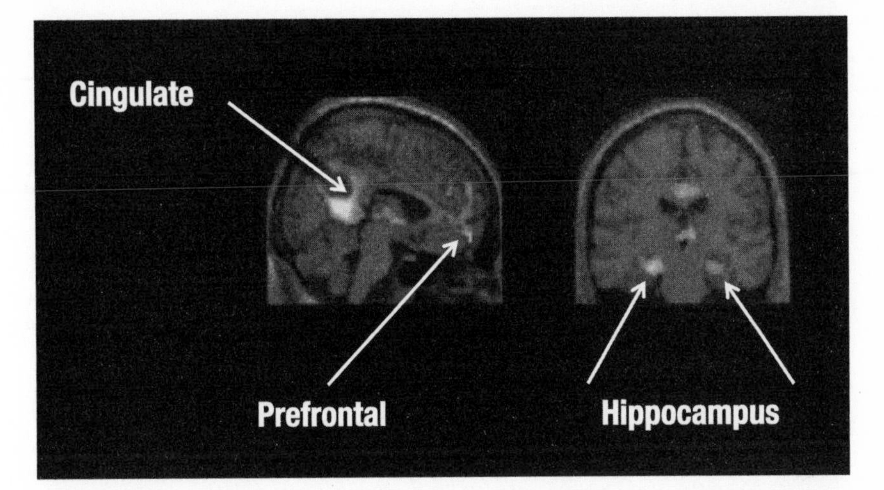

Figure 4: Brain activity while retrieving autobiographical memory. Many regions have increased activity in limbic structures, including the prefrontal cortex, hippocampus, and posterior cingulate cortex. Patients with a damaged left hippocampus suffer from severe autobiographical memory loss.

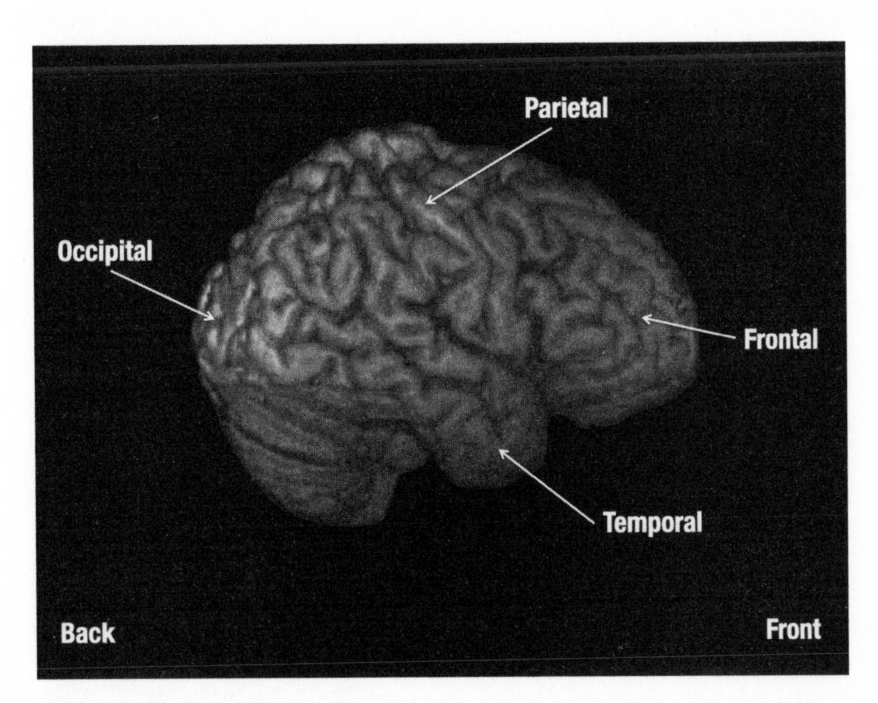

Figure 5: The major brain regions

of Iowa describe a woman with severe memory loss who lived in a world with a history lasting seconds to minutes. She could recall events from her youth and adulthood that happened before a disease destroyed the brain structures critical to her forming new memory. Her entire world was a narrow window slit, in which she constantly forgot what had happened minutes earlier. Yet she had a sense of self, acknowledged happy feelings, expressed sympathy for others, and was aware of her body. I have observed many patients with Alzheimer's whose memory loss is so severe that they don't know where they are in time or space and don't acknowledge the most basic autobiographical data, yet they do consistently respond to their names. As our memory goes, knowing ourselves as individuals seems to be one of the last human qualities that endures.

If we lose self and with it past memories in spiritual experience, I don't think it is from a complete shutting down of the memory process. After all, the experience itself becomes one of our strongest memories. Instead, the processes in our brains that have to do memory *retrieval* shut down. Forgetting who we are in the moment can help lift us toward transcendence.

Da Vinci versus Picasso

Neurologists view the human mind and brain in terms alien to nearly everyone else, including other physicians. Because of our training and the kinds of cases we treat, we are forced to see the self in ways that are often startlingly counterintuitive. In this respect neurologists are much like astronomers who conceive of a massive universe with time, forces, and distances that we can never directly experience, feel, or see. These scientists work with mathematical reasoning and logic disconnected from common sense, a kind of knowledge that isolates them and is often in direct violation of ordinary daily experience.

Similarly, neurologists inhabit domains isolated by their insights into the anatomy, chemistry, and physiology of an organ whose depths may be as unfathomable as the astronomer's void above. The neurologist's world routinely strips away the illusion that self-identity is a seamless, integrated whole. Instead, neurologists are confronted with overwhelming spectacles of subjective experience often fragmented into contradictory components.

When I first laid eyes on the *Mona Lisa* in the Louvre in Paris, for example, I was drawn to Leonardo da Vinci's small masterpiece by the famous smile and struck by how the face before me appeared unified and immediately recognizable. Yet I knew that behind that familiar face was an illusion. I thought of Picasso's portrait of Dora Maar, who was the artist's muse, model, and lover. Picasso's fractured perspective creates facial features that have seemingly unnatural relationships and proportions. Most of us find da Vinci's *Mona Lisa* much closer to the way we normally perceive the world than Dora's disjointed planes. But the brain makes us who we are from a jumble of components that are fragmented and distributed through the cortex and thalamus in a way that is analogous to the way Picasso sometimes painted.

My Puppet Legs

When I was much younger I underwent elective surgery and had firsthand experience with the way the self can fracture and fragment. The body's longest nerves, those to the feet, are often the most vulnerable to injury. Diabetes is a cause for this very common problem (I see cases of this type of damage from diabetes all the time). Patients with nerve damage in the feet typically report they are "walking on blocks." Sensation from the feet can't make its way to the brain because nerve damage blocks the connection.

As a postsurgery patient, I had a similar experience. I awoke in an unfamiliar recovery room with no memory of the surgery or how I got there, thanks to medications that had not completely worn off. So I wouldn't be alarmed, the recovery room nurse told me that my spinal anesthetic was still working and my legs were paralyzed. My fingertips confirmed that my legs were numb, conveying to my brain no feeling that they existed. This was not troubling since I was groggy and had no desire to do anything but lie there quietly. Soon, however, the nurse decided to take me to my room. As the gurney stopped before a door, I was cautioned that the legs, although stronger, remained weak and would not support my weight when I transferred into bed. I decided to test my condition. When I tried to lift the right leg and extend the knee, the limb flailed wildly in every direction except where my brain told it to go. The left leg was no better. I had little sensation of the legs moving and could not feel where they were in space. This gave me the strange sense that I was disconnected from the waist down. I could see the legs, but they felt as if they were strung to some other puppet master.

Our muscles must contract and relax in a tightly coordinated way to smoothly move our limbs, whether we're walking across the room or racing downhill on skis. After my operation, nerves normally conveying my leg sensation or connecting to my leg muscles were blocked. My legs were no longer agents of my will.

I often see cases where it's clear that the self is similarly nebulous. We can have a hard time determining what happens in the brain to make the self lose cohesion and what parts of the brain are necessary to reconstitute it. The self, like consciousness, can fragment. And when we look closely, it may be that a little piece of the self remains even in the most powerful experiences of self-transcendence. After we delve into the way the brain constructs the self, we see that the complete loss of self, even in the most powerful of mystical experiences, is a dubious proposition.

I know of four examples where the self is lost or found that provide clues to how the brain puts us together. We are going to start small and focus on an arm. In three of these cases, the brain was injured. In each case you can see a way we might distort the self during a spiritual experience.

The Alien Arm

My legs returned to me after my surgery, and I was not so much distressed as intrigued by this disconnected feeling, knowing it was temporary. I can only faintly imagine what it is like for my patients, like Steve, who are permanently paralyzed.

Steve was startled when he awoke in his rural Kentucky home at 3 A.M. to go to the bathroom, got out of bed, and fell to the floor, unable to walk. In the local ER, a computerized tomography (or CT) scan of his head showed traces of an early stroke, so he was moved to the University Hospital, which is where my colleagues first saw him. It was immediately clear that something very unusual was wrong with Steve. He professed weakness in his left arm and leg but they tested fine. More significantly, he ignored his left arm, glaringly unaware of its presence. Even stranger, when asked to move his right arm, his left arm would involuntarily mirror his right arm's movements in a seemingly irresistible imitation of what his right arm was doing. Steve could not will his left arm to move—it moved of its own volition! If he tried to use it for tasks such as eating, it would not obey his commands, although his arm had normal strength and sensation, such as touch, that was conveyed to his brain. When asked why his arm was out of control, he said: "It does what it wants to do, and not what I want it to do. I try to do things with it, but it just does not want to work." His left arm had developed a will of its own, become foreign, alien to his self. It was similar to Thing, the disembodied hand from the Addams Family.

Alien limb, as it's known, takes many forms. Sometimes the alien limb (which is almost always on the left) is at odds with its opposite. The alien limb pulls up trousers that the normal limb is trying to remove. Patients with this syndrome can be startled because they think someone else is in the room when their alien limb opens a door, removes the sheets from their bed, or comes suddenly into view. The alien limb can purposefully grasp at anything in its reach, almost as if it is magnetically attracted. Its owner disavows responsibility; the limb, the owner says, disobeys, defying his or her will. In some patients, the "good" arm is kept busy, restraining the alien one from unwanted, possibly harmful acts. I have heard of cases where an alien hand grabs the steering wheel and tries to make a sharp turn, causing the car to veer wildly. The driver's normal hand has to restrain the alien one to prevent a crash. Though rare, this condition is well documented by neurologists, and it is always caused by damage to the cortex. A limb has gone amok—body and self are divided.

Underlying the alien limb is a disconnection from consciousness and other parts of self; the specialized nerve circuitry controlling the arm is cut off from other brain areas, leaving the disconnected area isolated but functioning in an often bizarre or autonomous manner. Other disconnections may occur. A patient who suffers a stroke that injures the left side of the brain may retain the ability to read but not write. If the brain region devoted to left hand sensation is disconnected from language regions, when the patient closes his eyes he can use a toothbrush in his left hand but will not be able to name the object until he opens his eyes and the other (left) half of the brain, containing our language ability, actually sees it.

In Steve's case, his stroke was in two locales: the right parietal lobe, which is dominant for identifying ourselves in the world and enabled him to identify his left arm as his own, and the corpus callosum, the large bundle of nerves that connects the right and left sides of the brain.

I first saw Steve three days after his stroke. By this time, he could control his left arm: he could dress himself and perform other simple tasks. He still mostly ignored the arm, but I expected this to improve. His stroke wasn't as severe as other cases I've seen.

I cared for Robert, another patient, also with a right parietal stroke, who was sent to me for testing of the nerves in his left arm. When I asked him to move his arm, he replied: "I've tried to tell them that the arm is not mine."

"Whose is it then?" I asked.

"I don't know," he replied. "But it has something to do with my stroke. It's either mine, and I've had a stroke, or I'm nuts."

Both Steve and Robert had these glimmers of insight about their condition. They shifted between disowning their left arm and realizing the bizarreness of doing so. Neither could quite yet reconcile the two incompatible ideas. This awkwardness was a good sign for recovery.

The immense turmoil caused by Steve's stroke led, as far as I could tell, to no particular transcendent insight or feelings. This suggests that simply turning off the right parietal lobe and its connections isn't what is happening during spiritual experience.

A Magical Touch

Magicians and psychologists have long used illusions and trickery to serve their professional interests. Research psychologists have used one of these tricks to illustrate how the brain uses vision, touch, and where our limbs are in space to build the body's sense of itself. Matthew Botvinick and Jonathan Cohen, psychologists at the University of Pittsburgh, seated an unwitting subject at a table. One of the psychologists told him to place his left arm behind a screen so the arm was shielded from his view. The psychologist placed a lifelike rubber arm in front of the screen and told the subject to fix his eyes on the fake arm while each arm, the rubber one and the real one behind the screen, was lightly brushed with the same rhythm. After a while, the psychologist stopped brushing the real arm, only brushing the rubber one, and the subject reported that he could feel that.

Later, the researcher told him to close his eyes and move his right hand near his left. The subject's right hand consistently ended up closer to the rubber left hand than his real hand. The subject felt that the rubber hand, rather than his real hand, belonged to him.

The brain's ability to incorporate an inanimate object like a rubber arm into the body's schema graphically illustrates the fragility of self, the fragility of the brain processes that organize our perceptions. But this breakdown did not coincide with a spiritual experience—during which loss of self can be lightning-fast. It took several minutes for the experimenters to disconnect the real left arm and substitute the false one. These are not conditions that are likely to be found in a real-life setting. We don't know how the brain took on this artificial appendage, but if this process could be rapidly expanded to include all the world around us, even the entire universe—not just a rubber limb—it might show us something about what happens in our brain during a spiritual experience. Either way, the rubber arm trick reveals the power of the brain to imperceptibly rearrange the self.

Chasing Out Phantoms in the Brain

S. Weir Mitchell, a novelist and friend of William James, was one of America's first neurologists. Mitchell was a surgeon during the Civil War and performed many limb amputations. As they recovered, his patients told him about the persistent "sensory ghosts" that replaced limbs they had lost. Mitchell, who wrote exhaustively about these war injuries, is credited with providing the first extensive medical descriptions of what became known as phantom limbs.

The sensations in phantom limbs can amount to vexing lifelong pain so excruciating that it drives physicians and patients to desperate measures, including extreme surgical remedies, such as destroying parts of the brain. Only in the 1980s did neuroscientist V. S. Ramachandran finally, and quite cleverly, devise a remarkably simple yet effective treatment for phantom limb pain, based on what the phantoms really are—a distortion in the way the brain maps our sensations.

Assembling the map relies on the fact that, contrary to what you might think, the brain itself is insensitive to touch and pain. Probes can be placed on top, or inserted painlessly into the deepest brain regions since the brain has no physical sensation of itself. Headaches come from nerve endings in the blood vessels and protective brain coverings, not the brain itself.

The brain's insensitivity allowed Canadian neurosurgeon Wilder Penfield to stimulate it, hoping to distinguish the normal brain from tumors and scar tissue that caused seizures in his patients. He recorded where patients felt sensations during stimulation or what part of their bodies reacted to it. By repeating these tests hundreds of times, Penfield was able to map where in the brain our body parts are represented. It is interesting that the body parts needed to speak, compose facial expressions, make intricate hand movements, or convey complex sensation—like fingertips and lips—take up a larger portion of the brain. Our evolution has substantially depended on these very human functions—they get lots of real estate in our brain.

At first it was thought that once these maps were formed they remained unchanged, but it turned out that they drift like the continents and can reorganize.

When a patient loses a limb, the brain reorganizes its sensation map. And the patient can end up feeling things in a limb that doesn't actually exist. How can a doctor treat such a bizarre malady?

One day, Phillip, a man in his thirties, walked into Ramachandran's office, desperate for help. His phantom left arm had been

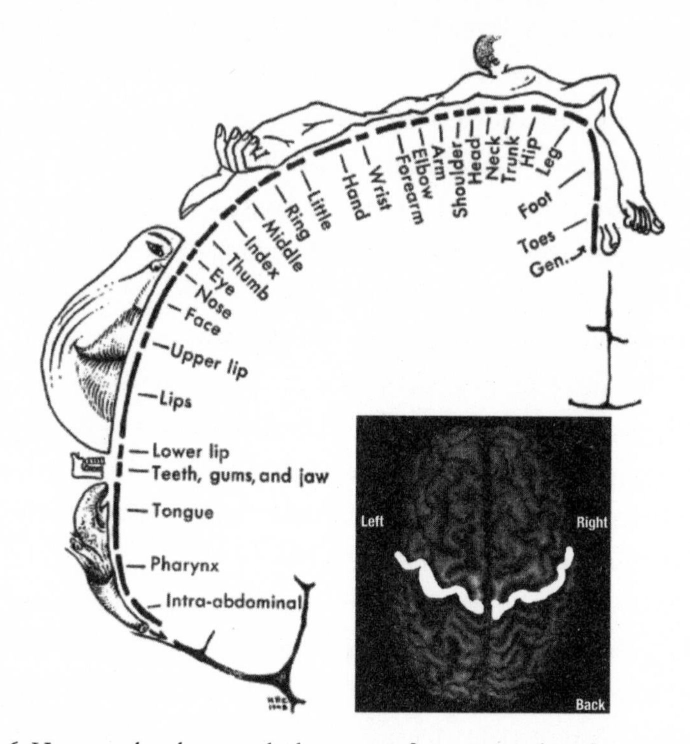

Figure 6: Homunculus showing the brain map for sensation based on brain stimulation by Wilder Penfield. From above (inset), white regions show where the sensory homunculus is located.

painfully frozen in an awkward position since it was amputated several years earlier. Ramachandran was ready. He asked Phillip to sit before a box the neuroscientist had designed that contained a simple dime-store mirror. Holes were cut for Phillip's right arm and where his missing left arm should have been. The mirror was situated so that when Phillip "inserted" his phantom left arm, the mirror image of his right arm appeared as his painful phantom left arm. Phillip exclaimed: "This is unbelievable! My left arm is plugged in again." For the first time in years he could feel his left elbow and wrist move. But he could only feel the "movement" if he looked into the box. When he closed his eyes or took his right arm out of the box, the phantom limb returned to its frozen position.

Ramachandran had Phillip take the box and mirror home and use it for ten minutes each day. After several weeks, Phillip reported that his phantom limb had vanished. No longer did he have a painfully contorted arm. Phillip had undergone the first amputation of a phantom limb! The mirror image of his right arm had resurrected the amputated left arm, allowing Phillip's brain to reorganize its map and return to a more normal condition. The technology for this treatment was certainly available to Weir Mitchell for his Civil War veterans–if he only he had known more about how the brain works.

Although the body map within the brain is malleable, it doesn't seem malleable enough in this way for spiritual experience. It takes considerable effort over weeks to redraw a real brain map, less time than moving tectonic plates, but still it doesn't seem likely that the brain map to our body is going to shrink fast enough to lose the self in spiritual experiences that last sometimes only seconds. The brain's power to create not only what we experience as objective reality out there, but also our own flesh and blood, is astonishing. But the time frame must match that of our spiritual experiences.

A Spare Arm

Amputees are not the only ones with phantom limbs. A brainstem stroke can also cause spare limbs to sprout. Dr. Hideaki Tanaka and his colleagues in Japan cared for a forty-seven-year-old woman who had suffered a devastating brainstem stroke. She could think normally with her cortex, but she was completely paralyzed except for the ability to move her eyes up and down. This horrific condition is known as "locked-in" syndrome. The stroke disconnected her brain's body map. When she recovered sufficiently to be able to describe what she felt while she was paralyzed, she said an extra arm had formed on her left side. She could move it, although at times it seemed to have a mind of its own and it felt as though it were trying to strangle her. Sometimes a motionless third leg would appear just inside her left leg. These spare limbs disappeared as she recovered.

Extra limbs are different from phantom limbs. How the brain makes them is unknown. Perhaps the brain map of the woman's left arm, cut off from her lower body, somehow reorganized and spread into areas that were not arm regions before the stroke.

All the examples above illustrate how our sense of self can be disrupted. These disruptions come not just from the cerebral cortex, but from nerves all over the body, our legs, arms—all over. Our sense of self is fragile. If my arm can be alien to me or seem to still exist even after it is removed, if a rubber appendage can be my flesh and blood, if I can sprout new limbs from my body—can I trust the seeming reality of an out-of-body experience? Are the revelatory changes in one's sense of self during spiritual experiences just kindred illusions? If the self could be thought of as a figment of our imaginations, could the same be said of spiritual truth?

The "Me" in My Brain

Self-awareness is conscious knowledge of one's own character, feelings, motives, and desires. Just as we do not know how and what part of the brain it takes to be minimally conscious, we are uncertain what brain content we need in order to be the "me" of our human selves, although we do have some ideas. Many of these ideas originate with the Victorian epileptic seizure researcher mentioned earlier, John Hughlings Jackson.

Jackson's keen interest in mind, brain, and self combined meticulous observations of his neurological patients with what, in his day, was the new theory of evolution. Jackson suspected that our frontal lobes enabled us to detect ourselves, and, rightfully, he reasoned that we should look for the human self in the most recently evolved areas of the brain. He called the foremost frontal lobe regions the "prefrontal," where "mind" originated, facilitating the complex coordination of other self features that arise from the cerebral cortex. He called disturbances of the prefrontal self "dreaming states," which were most often caused by epilepsy.

It has long been known that injury to the frontal lobes has a profound impact on personality, an essential component of the self. Permanent and irreversible personality changes ensue if the brain injury is permanent. Those who suffer brain degeneration, physical injury, or tumors in this region often have dramatic changes in dress, political philosophy, and religion. A woman with a taste for elegant designer apparel may suddenly prefer cheap, gaudy clothes. The puritanical can become sexually experimental; the frugal, spendthrift; the scrupulously honest, compulsive liars and cheats. Religious faith and behavior can radically change as well.

One of the most famous cases of this type was railway foreman Phineas Gage. On September 13, 1848, Gage was busy leveling ground to lay new track near Cavendish, Vermont. Holes drilled in rock were filled with blasting powder. Gage tamped the powder tight with an iron bar and accidentally sparked an explosion that rocketed the bar upward through his skull: it entered just below his left eye and exited at the top of the head. Miraculously, he was only momentarily stunned. It was evident that the rod had missed all the vital structures. Yet once his wound healed, his damaged frontal lobes transformed him from a hardworking, socially responsible, law abiding, and reliable person into someone who was notoriously capricious, profane, lazy, and irresponsible. Memory and speech and other brain functions were unchanged—it was his *personality* that was irrevocably transformed.

Gage sustained injury to the prefrontal cortex on both sides of the brain. We now know that this area is important for rational

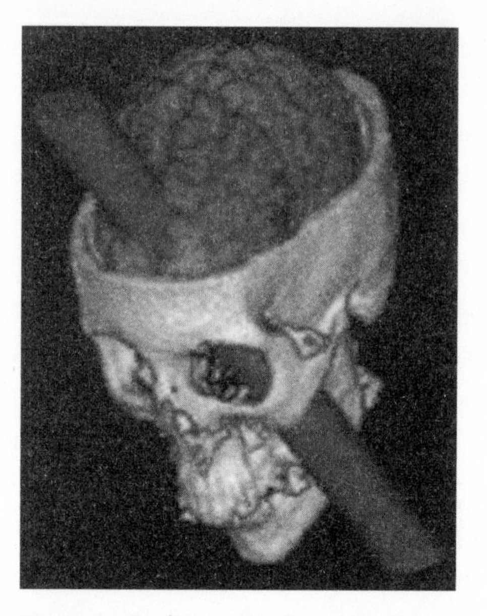

Figure 7a: Rod as it passed through Gage's brain

Figure 7b: Gage holding the famous rod

decision-making and processing emotion. Reflecting upon oneself is a necessary and perhaps first step in achieving self-awareness or, as James said, knowing there is a knower. In one MRI study, subjects were asked a standard series of true or false questions beginning with "I" or "my," such as: "I get angry easily," "I often forget things," "I can be trusted." When their scans were compared, all the subjects had activated prefrontal regions. There was activity in the posterior cingulate region as well.

The posterior cingulate is an important limbic area for autobiographical memory, guiding our responses to a familiar face or voice, and perceiving and processing emotional stimuli. In those moments when we reflect upon ourselves, our MRI scan looks remarkably like our MRI scan when we retrieve autobiographical memories. The posterior cingulate may also be important for generating admiration and compassion. It must be important in some types of spiritual experience, where extreme admiration or worship is involved or where one feels intense compassion.

We know the prefrontal and posterior cingulate regions are important for the self, but that still does not conclusively determine where the "me" appears in the brain.

Two Unequal Halves

The cerebral cortical mantle is a rough sphere overlying the rest of the brain, with right and left hemispheres, halves connected by a very dense band of nerve fibers, the corpus callosum (and lesser tracts). These two hemispheres communicate information and knowledge from one side to the other, making our minds whole.

In the nineteenth century phrenology was popular. Phrenologists argued that certain brain regions caused specific personality traits and enlarged to form bulges in the overlying skull. Thus it was thought that palpating the contour of someone's skull could give special insight into a person's weak and strong qualities. Formulating his ideas in this atmosphere of pseudo science, Arthur Wigan in Great Britain proposed that the two cerebral hemispheres were not homogeneous halves; rather, the brain was two distinct entities, each hemisphere processing an independent and complete mind. His phrenology-based reasoning was wrong but his conclusion correct—a conclusion that resonates with our current notions about human consciousness and self.

I am fascinated that there could be more than one conscious self within a single brain. Is each capable of a spiritual experience? If so, what kinds of spiritual experience would each side have? How would they differ? Might one side of the brain feel itself blessed while the other thought it was damned?

Whether we could have two independent minds in one brain is a question that could not be addressed until the middle of the twentieth century, when the right and left brains could be physically disconnected and isolated from each other. Modern split-brain research began in earnest when a World War II paratrooper's brain injury caused life-threatening seizures that could not be controlled by medications. It was decided by neurosurgeons Philip Vogel and Joseph Bogen at Loma Linda University in Los Angeles to sever the paratrooper's corpus callosum between the right and left cortical hemispheres. The surgery succeeded, greatly reducing the patient's seizures in number and viciousness. The doctors built on their success and similarly treated several other patients, who were then scientifically studied by Roger Sperry and his colleagues, including Michael Gazzaniga at the California Institute of Technology.

Sperry received the Nobel Prize for his insights into the self. He found that usually after this type of surgery little seems amiss. There are no apparent deficits of thinking, memory, mood, or behavior. Split brain patients use their hands and feet in a coordinated manner. Both sides of the brain draw on nearly the same autobiographical memories and create new ones with common experiences. Each side independently recognizes the patient's own reflection. In short, the sense of self seems remarkably unaffected. This is not surprising: most of the time the two halves of our brain are in synchrony. Information from the surrounding external world and the internal world comes to the right and left brain simultaneously.

When the connection between right and left brain is completely severed, the hemispheres continue to share the same conscious state because they share the same arousal system. In other words, one side is not awake and the other asleep. The brainstem arousal system projects upward, going to both the right and left sides before reaching the thalamus and cortex. However, attention to the outside world, the attentional "spotlight," now exists independently in both hemispheres.

Although not obvious, after the two brains are split, there are subtle changes. The first clues that two minds could be in one brain came from observing the brain halves at cross-purposes, just as happened with the alien hand. Shortly after surgery, a young boy was seen dressing, pulling his pants down with his right hand while pulling them up with the left. The right brain controlling the left hand did not know what the left brain controlling the right hand was doing. This was not well understood until neuroscientists examined patients under exacting experimental conditions where information was given only to one side of the brain. These experiments, by the way, gave rise to the popular expression of being "right" or "left" brained.

Although today neurosurgeons rarely split the brain, some people with seizures that cannot be controlled with medication have surgery to remove the brain section that is the genesis of their abnormal electrical activity. Before operating, the surgeon needs to carefully look at the brain regions for language and memory. A catheter is placed in the right and left carotid arteries that lead to the brain. Amobarbital, a powerful sedative, is injected, briefly anesthetizing one side and then the other side of the brain. A neuropsychologist asks questions and shows pictures and objects to the side of the brain that is awake, probing its language and memory. This tells us which side of the brain has the most memory and how much language is in each hemisphere. The results can surprise both doctor and patient.

One of my neuropsychology collaborators in near-death research, Dr. Frederick Schmitt, has directed many such tests. Two of his patients were particularly unusual. When their right hemispheres were injected, their left hemispheres came alive in unexpected ways. Quiet and reserved patients became sexually boisterous, talking about carnal matters and trying to hook the medical personnel up with one another. Afterward, when their right hemisphere returned, they were embarrassed and contrite.

After their brain is split, it is not apparent at first that the patients might have a separate consciousness in each side of the brain. For most people, most of the time (split brain or not), the right hemisphere is silent. In the vast majority of us nearly all language ability resides in the left hemisphere, leaving the right hemisphere with little capacity to read and write. This means that many experiences of the right hemisphere are contained, hidden from the left hemisphere and the outside world, if the corpus callosum is split.

Paul is science's most famous split brain patient because his right brain had the capability to communicate with the outside world. As part of the postoperative routine, he was given a series of tests by Michael Gazzaniga, then at Cornell Medical Center in New York, teamed up with Joseph Le Doux and Donald Wilson, to probe the language ability of his right hemisphere; to their delight, researchers found they were able to communicate directly with Paul's right hemisphere.

When the researchers began examining Paul by presenting information, such as written questions, directed only to his right hemisphere, his left hand was able to spell out his responses with Scrabble tiles. The first question they asked was "Who are you?" "Paul," spelled his left hand to their great excitement. This was the first time anyone had knowingly and directly communicated with the right hemisphere.

Figure 8: Only Paul's right brain sees the full question of "Who are you?" which the right hemisphere correctly answers "Paul."

Soon Paul was naming his favorite TV personality, the Fonz from *Happy Days*. He was then given a series of items to rank as being "good or bad." His right hemisphere had a darker, more negative view when it came to war, sex, his mother, and even himself. In contrast, money and his girlfriend were on the "good" side of the scale for both hemispheres.

When asked to rate what he "liked or disliked," his hemispheres were in accord. Both liked TV, sex, school, church, and the Fonz. Only "dope" was discordant: the right brain liked it while the left disliked it "very much." As researchers probed, other differences emerged. Paul's right hemisphere wanted him to become a race car driver; his left, a draftsman.

Paul's hemispheres probably shared emotions, because the connection between right and left emotional (limbic) structures was intact. The mood of one hemisphere may trigger the same mood in the other hemisphere, even when the second hemisphere is unaware of what provoked the new mood.

The researchers' work with Paul demonstrates that both hemispheres, independent of each other, have a sense of self, a feeling that they are the agents and owners of their own separate experience, the individual experience of selfhood. The right hemisphere has an understanding of the future, with its own aspirations, likes, dislikes, and opinions of good and bad. Though the hemispheres draw upon the same autobiographical memories and exist in the same emotional milieu, there is much to indicate in Paul that the two sides of the brain, once split, are independent consciousnesses. What we see is split consciousness and self, with the two unequal halves, once whole and now living side by side.

My neurological instincts tell me that two different minds from the two discrete hemispheres of the brain, which have very different attributes, must lead to different expressions of the sacred. Each hemisphere contains a separate self with its own likes and dislikes, its own moral shadings. It may also have its own ideas and experiences of spiritual insight or truth, different thoughts and feelings about the holy and the divine.

Two Minds in One Brain

Again, could one hemisphere feel blessed while the other felt damned? Extending this line of thought: could one hemisphere be capable of not only wallowing in profane imaginings but committing immoral acts, while the other had absolutely no responsibility for such behavior? Or could one side be wracked with guilt while the other self-righteously trumpeted virtue? It seems we see that kind of discordant behavior often today in religious personalities and politicians.

Early in evolution the vertebrate brain developed different right and left sides. This allowed the brain to develop more highly specialized regions than if both sides did all things equally well. In the human brain, this cortical specialization is the most refined of any species'. While the left brain handles language, the right is better at recognizing faces and has superior visual-spatial perception. You do not want to count on your left hemisphere to navigate you around the house, let alone in unfamiliar territory.

The halves also think very differently. The left hemisphere excels at organization and problem solving; the right struggles with these tasks. The right maintains a veridical (objective, straightforward, unembellished) record of experience and appreciates important aspects of causality. The left generates hypotheses, interprets patterns, and makes associations (even when none exist!). It creates stories, constructs theories, seeks explanations, and gives us the "why" for our experiences. Accuracy is not the primary concern of the left hemisphere; rather, its priority is weaving experience into a comprehensible or at least explainable whole.

In other words, the left hemisphere not only talks—it also makes things up! The left hemisphere is driven to explain behaviors arising from the right hemisphere. In one patient, a nude picture was shown to her right hemisphere, causing her to giggle. When asked why she laughed, her left hemisphere, which had not seen the picture, replied: "That's a funny machine." Her left hemisphere may have shared the right's emotion on seeing the picture, but the left brain is fond of storytelling and could very well have confabulated her explanation of why she found herself giggling.

It is tempting to speculate that as primates developed their interpretive, language-rich left hemisphere, they began interpreting their observations of nature in a supernatural context. It could be that the left brain first brought us our gods; it certainly allowed us to talk about them. In fabricating a running narrative of our emotions, thoughts, memories, and dreams, the speaking left hemisphere may be critical for creating the illusion that we are a whole self. It may be responsible for making the self explicit. The ultimate interpretation could be: Who is doing the interpretation? Me!

Even when the two halves of the brain are split in surgery, there is usually some remnant of communication between the hemispheres. Although this limits some of the conclusions that we can draw, it does not take away from the fact that two separate consciousnesses are created by a split brain. Consciousness is doubled, but consciousness content is different between the right and left brains because the specialized circuits are not equally distributed between the hemispheres. In all of us, the "me" is distributed throughout our brains.

We don't know whether Paul's right hemisphere had a self in its consciousness only because it also had language. Does language make the self or does language only make the self detectable? Considerable evidence tells us that the right hemisphere can, on its own, be a human self without language.

Every neurologist has witnessed the devastating strokes that destroy the entire left hemisphere, leaving patients completely paralyzed on their right side, unable to speak, understand others, or think in language. Only their right cerebral cortex remains, and these patients are much more thoroughly disconnected than after split-brain surgery. Through smiles and anguish, patients recognize themselves and their loved ones by face and voice, and they nonverbally respond to the world around them in appropriate ways. They remain, to both me as their doctor and their families, a familiar person—indubitably mother, father, child.

Ironically, although the left hemisphere tells others who we are, it is our right hemisphere that contains much of the circuitry important for us to recognize our own face. The capacity to reflect on one's own self is important to self-awareness and likely part of some spiritual experiences as well. The medial prefrontal (later we will find it important to bliss in spiritual experience) and posterior cingulate regions containing the machinery for self-reflection—when I ponder whether I am a good friend, or the disciple worries if he is making progress in the eyes of his master—are used on both the brain's left and right sides. A self, a sense of agency and ownership, exists within the right, silent hemisphere. But it is clear that what you think is you, the decision maker, is distributed through your right and left brains; the process of bringing these parts together is what creates the whole living self. Usually.

I Do Not Exist!

On June 28, 1880, Dr. Jules Cotard delivered a lecture in Paris and presented a case that is still startling today. A forty-three-year-old woman had the unshakable belief that she had lost her brain and other organs; she was "eternal," and "neither God nor the devil existed." In other words, she thought she was dead! Cotard attributed this delusion to severe depression (delusions are pathological beliefs that remain fixed in the face of clear contradictory evidence).

Only a few cases of Cotard's syndrome have come to medical attention over the years. Even seasoned neurologists are stunned when a seemingly rational patient, speaking normally, knowing where he is in time and space, capable of following complex commands, insists he is dead. Such a case was brought to my attention by one of my friends and colleagues, Dr. E. Wayne Massey at Duke University.

Earl, a fifty-eight-year-old man, was brought to the hospital because of a brief spell when he couldn't speak clearly. A few weeks earlier, he had had a major operation to repair a weakened bulge in his aorta. After the surgery he had become depressed and frightened about dying and his thinking "was in a fog." He had been put on

Coumadin, a blood thinner, and his doctors were naturally concerned that he could have suffered bleeding into the brain. He was extensively evaluated with MRI scans and other tests. Each was reassuring. But what Earl told Massey shocked him. A few days before, he had been sitting on the toilet when it suddenly occurred to him that he was dead. Not just a little dead, but completely dead! Over the next few days he tried desperately to convince his wife and others that this was, indeed, the case. For obvious reasons, they did not believe him.

In the hospital, Massey could find nothing wrong with Earl's brain or nervous system. He came to the same conclusion as Cotard: the patient was psychotically depressed. Earl improved over the ensuing days, but the mystery of what had led him to believe he was dead remained.

Cotard's syndrome patients think they are dead and that their bodies are decomposing or they are walking around in an afterworld. In its fullest expression, the patient is convinced that he or she does not exist, in the face of clear contradictory evidence. Cotard's is plenty odd in and of itself, but it is often also accompanied by Capgras delusion, a condition in which a person believes that impostors have replaced loved ones or familiar people. Only the patient detects the crafty substitution that has fooled everyone else. Rarely has a stranger or near stranger been substituted: the impostor is nearly always someone important to the patient.

It has been argued that the right hemisphere in a patient with Capgras fails to recognize the emotional value of the person. The left hemisphere then unleashes a "creative narrator" who confabulates false explanations.

Cotard's syndrome has been attributed to many causes, all of them from widespread injuries such as concussion, seizures, a temporary interruption of blood flow to the brain, schizophrenia, or Parkinson's disease. The brain dysfunction that causes Cotard's is not clear even in the presence of discrete seizures or brain injury. No part of the brain can be found pointing to Cotard's syndrome, a region where we can doubt we exist. If such a region could be identified, it might also be where the brain *affirms* our existence. The many questions about Cotard's will probably remain because it is so rare and fleeting. But knowing that Cotard's exists should give us pause. The power our brain has over our most important assumptions is awesome. We need to keep this firmly in mind as we examine spiritual experiences such as the mystical and near-death. The brain is perfectly capable of creating experiences that are utterly convincing and are often described as "realer than real," as though the person experiencing them had stepped through a doorway into a deeper, truer apprehension of the nature of the universe and the meaning of life.

Nothing As It Seems

Realizing that self is a separate process contained in the process of consciousness may not feel like a revelation after reading these last two chapters. Yet in ways subtle and obvious, the self and consciousness have been confused with each other in neuroscience, philosophy, and psychology for years. They are especially confused in writings about spiritual experience. We must keep the processes of self and consciousness as separate as we can in our minds while we explore spiritual experience.

Spiritual experiences can fragment consciousness and reverberate in a fragmented self. Some spiritual experiences transform self as their only quality, whereas others—mystical experiences, for example—can take away your sense of self and leave you with something miraculously new.

The way the self is put together in the brain is a fragile process. Philosopher and neuroscientist Thomas Metzinger makes the case that the first person perspective arises from a process that is transparent to itself. You cannot perceive your self process, but you perceive *with* the process. He calls this transparency "a special form of darkness." The brain has no inner eye watching how nerve clusters are weaving together to make the cloth of the whole self. Although we cannot perceive our own self process, we can understand that process to some degree in others

Within some people's lives there arises a special form of enlightenment that always touches on the self. The self can be embraced, lost, and changed—all in the same experience. It often seems to people in the midst of this kind of experience that the self is standing in the doorway to the afterlife, heaven, a place of beauty and serenity . . . where they encounter the dead who have been important in their lives; people they have loved or who have loved them. The voyagers returning from these experiences say that their visionary journey was "the most real thing I have ever felt."

Let's listen to people who have had these near-death experiences, now that we have tuned our ears to the brain, consciousness, and self. After all, their stories changed the course of my life.

PART TWO

AT THE DOORWAY

4

THE VARIETIES OF NEAR-DEATH EXPERIENCE

TELLING STORIES

"It is natural that those who personally have traversed such an experience should carry away a feeling of its being a miracle rather than a natural process."

—William James, *The Varieties of Religious Experience*

Throughout human history people have told astonishing stories about returning from near death. It seems reasonable to assume that near-death experiences were at least partially responsible for the River Styx in ancient Greek mythology, the elaborate preparations of mummies for the afterlife in Egyptian tombs, the Tibetan Book of the Dead, and Christian images of an afterlife.

The part of our brain responsible for these stories extends backward to our primal origins, thousands of millennia before our brains developed language and we could tell the stories. My research unites our brain's survival reflexes with other ancient regions where we can reasonably say our "god impulse" resides; an area responsible for dreaming and emotions.

By moving the spiritual trigger from the cortex to the brainstem, I am suggesting a whole new line of investigating spiritual origins. This trigger could catalyze many types of spiritual experiences that have arisen in all epochs and cultures. As we look at the stories of near-death experiences, we will see how remarkably consistent they are, and how they point to our common biology.

To grasp the brain's role in these ancient stories, we must know about the stories themselves. The near-death experience has become a staple of books, movies, and talk shows. It's probably safe to say that millions of Americans consider these experiences as proof that an afterlife exists and that our consciousness or souls can separate from our bodies.

Skeptics, on the other hand, maintain that the spiritual dimensions of near-death experiences are drug-induced or illusory, the product of a blood-starved brain, wishful thinking, or both.

Yet no matter what we think of the veracity of near-death experiences, they have certainly shaped our view of dying. Going through a tunnel, being enveloped by "the light," leaving the physical body behind, meeting people we loved who are already dead—all of these are now in our culture almost stereotypical images of what is in store for us as we die.

And of course it is not only Americans who have near-death experiences. People across cultures report the sense of being dead, intense light, encountering dead relatives and spiritual beings, life review, approaching a border, and out-of-body experiences But in Japan, the uncrossed boundary is likely to be a brook or river. Tunnels are uncommon there, while here in the United States, they are ubiquitous. A "dead" person in India will be sent back into life if his or her name does not appear on a roster similar to the one belonging to Chitragupta, attendant to the Hindu king of the dead.

Some Americans have sighted Elvis during their near-death experiences, including Beverly, a banker's wife in the Midwest: "We really hit it off," Beverly is quoted as saying by Raymond Moody. "[Elvis] was in this place of intense, bright white light. He just came over to me and put his hand softly in mine and he said, 'Hi, Beverly, remember me?' Elvis sort of receded into the light and suddenly my father was there with me . . . gently telling me that I had to go back to finish my life, and that I would come again to that beautiful place when I died. So I felt myself being drawn backwards rapidly and the light receded, and then I felt a pop and I knew I was back in my body and that I would live."

Although a cross-cultural comparison is fascinating, there are strong limits to how much it tells about how the brain is working as a near-death experience occurs. An analogy might be the way different cultures experience hunger. All people feel and satisfy hunger. Yet comparing the means by which different societies obtain, prepare, and consume food tells us little about what nutrients are necessary for the body or the biochemistry of how the gut extracts those nutrients during digestion. So it is with near-death experiences—our common biology produces shared features, but each culture imparts its own distinct flavors.

A Common Occurrence

There is strong evidence that near-death experiences are relatively common—as many as 18 million Americans may have had one, according to a 1997 issue of *U.S. News & World Report*. Although common, until fairly recently these experiences have been treated as anomalies. Well-publicized celebrity accounts did much to change this. Elizabeth Taylor has had several, including an encounter with her dead husband Mike Todd: "I went to that tunnel, saw the white light, and Mike [Todd]. I said, 'Oh Mike, you're where I want to be.' And he said, 'No, Baby. You have to turn around and go back because there is something very important for you to do.' "

Being sent back with a new or renewed sense of purpose is a common feature of near-death experiences, as is a sense of death or an afterlife as a serene and happy place. Sharon Stone, who suffered a serious brain hemorrhage, described her near-death experience: "[I] went into the vortex of that white light [that was] very, very beautiful, and very comforting and very peaceful, and quiet and clean." Religious iconography is often featured. After a motorcycle accident, Gary Busey saw angels. "They don't look like what they look like on Christmas cards," he reported. He also saw "balls of light."

Near-death experiences are often accompanied by out-of-body experiences. Roy Horn, of the Las Vegas animal show Siegfried & Roy, almost died when his neck found its way into the jaws of his tiger, Montecore. While he was on the operating room table, Roy saw "a bank of white light, and then I saw all my beloved animals . . . For a moment I stepped out of my body."

Near-death experiences did not enjoy the attention or status they have today when William James wrote *The Varieties of Religious Experience*. He did, however, acknowledge the experience of our old friend Symonds, which occurred while Symonds was under the influence of chloroform and laughing gas during surgery. The more fulsome account of the incident is given by Symonds himself:

"[I had] a keen vision of what was going on in the room around me but no sensation of touch. I thought that I was near death; when, suddenly, my soul became aware of God . . . I felt him streaming in like light upon me . . . My whole consciousness seemed brought into one point of absolute conviction . . . I cannot describe the ecstasy I felt."

As Symonds returned to this life and his new sense of relation to God faded, he suffered unbearable disillusionment, and he flung himself on the floor before his frightened surgeons, covered in blood, crying out: "Why did you not kill me? Why would you not let me die?"

Once he regained his bearings, he was left with a burning question. Had he experienced "a delusion" or an "irrefragable certainty of God"?

That is, indeed, the question. But whether you believe that these experiences are doorways into a transcendent reality or clever illusions perpetrated by the brain, as we look at what may be happening in the brain at the time, we will get lost if we do not have a clear sense of what we mean by "near-death experience."

I want to examine what brain processes produce near-death experiences that are distinctly spiritual—life-changing events, rather than pallid oddities.

An Ambulance Takes a Detour

When I first began collecting stories of near-death experiences, I was struck as much by their differences as by their similarities. The spiritual content of each one seemed peculiar in at least one way.

Patrick came to me several years ago for help with his myasthenia gravis, a disease that weakens the connection between nerve and muscle. When his myasthenia acted up, Patrick would have nearly complete arm and leg paralysis and life-threatening difficulty swallowing and breathing. Patrick also had severe vascular disease, a narrowing of the arteries to his heart and brain. His greatest danger was dying from heart attack or stroke.

Fortunately, Patrick's vascular disease and myasthenia responded to aggressive treatment. After he was under my care for a number of years, he discovered my interest in near-death experience and told me the following story:

> I was forty-one, apparently in good health, working as an emergency medical technician in Knoxville, Tennessee. One day my team received an urgent radio call. The dispatcher sent us to a building that was on fire across town. I was driving the ambulance that day. I knew lives were at stake and floored it, weaving

> through traffic, siren blaring. I began to have chest pain, which I thought would go away at first. But it steadily got worse, and soon it was so severe that I had to pull over. I hooked myself up to the heart monitor in the back of the rig, and, sure enough, I saw that my EKG was real bad. I knew that I might be having a heart attack. Fortunately, my partner was with me and she gave me some nitroglycerine to put under my tongue and started an IV.
>
> A team of nurses and doctors met me at the door of the emergency department in a small community hospital and took my vital signs. My heartbeat was slow, and I was given medication to speed it up and stabilize its rhythm. I felt a tingling like needles all over my body. I was paralyzed and tried to get my doctor's attention but was unable to move or speak. My arms lay useless at my sides. Then I began to have this funny feeling. I lost the sense of having a body and found myself beginning to "levitate" above the gurney. When I first started floating upwards, I thought I had died. I tried to call out to the doctor but couldn't make a sound.
>
> Finally, I remember I managed to ask the doctor, "Doc, where are you?"
>
> "Right here," he replied.
>
> I raised my head and saw him beneath me. Higher and higher I rose, until I was able to survey the entire emergency department. I rose so high and close to the bright ceiling light that I worried I might burn myself. Soon I saw myself lying on the gurney, plain as day. My own face and mouth moved as I spoke with one of my EMT buddies. It seemed that I floated for a long time, but I have no idea how much time actually passed. The experience ended as I gently descended back to the gurney.

Patrick said that as he descended, he saw a nurse giving him medication to speed up his heartbeat, and he viewed his body from an adjacent storage room. He said he could see the nurse's station from that vantage point.

"Seeing" things during a near-death experience that are impossible to see—which has been explained as consciousness, a transcendent soul, or the self leaving the physical bounds of the body—makes this seem a spooky, supernatural story. But can we explain what happened to Patrick in terms of the brain alone?

As a neurologist, I was tantalized by a subtle fact in Patrick's account. When he first began "levitating," he did not "leave" his paralyzed body: he was specifically afraid the approaching light would *burn* him. This feeling of floating while prone, which began Patrick's out-of-body experience, is akin to what my patients with vertigo from inner ear problems report: they often feel as if their bodies are in motion even when they're still (similar to the feeling you have on a merry-go-round after it's stopped spinning). Patrick's body obviously didn't levitate upward toward the ceiling. He *felt* as if it did.

We shall examine in detail many of the major features of near-death experiences, including out-of-body states, in subsequent chapters. Suffice it to say here that along with the feeling that he was levitating, Patrick's brain created the illusion that his consciousness had left his body, just as the brain creates the illusion that we have a limb where none exists.

Patrick's EMT colleagues and doctor laughed with amusement when he was out of danger and recovering from his heart attack and told them what had just happened. Later, his doctor took Patrick's NDE more seriously, calling it a "remarkable experience."

Patrick obviously had a near-death experience, although some aspects that we have come to associate with NDEs were conspicuously absent. There was no feeling of peace or harmony with the universe. Patrick did not see a tunnel or a special light, meet dead relatives or spiritual beings, or feel a transcendent presence.

Was Patrick's experience spiritually barren? I don't think so. In the ambulance and on the gurney prior to his out-of-body experience, he knew he might die. As he began to "levitate," he said he felt he was about to "meet my maker." Asked if he believed an essential part of himself was actually floating in the room, he answered: "Lord, yes. I really do."

Still, Patrick's experience lacked the dramatic and obvious spiritual content typical of many near-death experiences—like the ones you might see presented on television.

A Deadly Allergy

Recently the actress Jane Seymour appeared on *Larry King Live* to discuss her brush with death. Seymour told King she had developed bronchitis when she was in Spain, filming the movie *Onassis: The Richest Man in the World.* The production team summoned a doctor, who injected her with antibiotics. Seymour immediately knew "something was wrong." Her throat constricted: she tried to call out for help but couldn't speak.

Seymour was in the throes of anaphylaxis, a life-threatening allergic reaction in which air passages in the lungs are cut off and a person is strangled by a swelling of the tongue and throat. This is a real medical emergency: if not treated within minutes by an adrenaline injection, blood pressure may drop low enough to cause shock, unconsciousness, and death.

"I remember I was panicking, and then I wasn't panicking," Seymour told King. "I was looking down at my body. I was half-naked. I had two huge syringes in my backside. I saw this white light. I had no pain. I had no tension, I just kind of looked, and then went, 'That's very strange. That's me. But that can't be me if I'm here.' And then I realized that I was out of my body, and that I was, you know, going to die. 'I'm not ready to go away. I want to get back in that body.

I have children I want to raise. And there's so much I want to do . . . I'm not ready to go.' "

Seymour's thoughts turned to "God, a higher power, whatever one wants to call it. I just said, 'Whoever you are, I will never deny your existence, just please let me get back in that body and I won't let you down. I will never let you down.' The next thing I knew I was in my body. It's very interesting because I was in control, but my body wasn't. My arms were flying. My legs were flying. There were two or three people there trying to hold me down."

The experience had a lasting impact on her. "I don't waste any time at all," she told King. "I spend more time with my children . . . I spend as much time as I possibly can doing things for other people."

Both Seymour and Patrick had near-death experiences where they were out-of-body, accompanied by low blood pressure, during a medical crisis. But Seymour saw a light and had the sense of a spiritual presence and divine intervention. Neither entered or even glimpsed another world, yet I don't think anyone would quibble with the argument that they both had a spiritual experience during a life-and-death crisis.

Near-Death Experience Defined

The near-death experiences described by Patrick and Seymour and the others mentioned earlier all fit into a profile developed by Raymond Moody, who compiled hundreds of cases.

The following table is based on Moody's accounts and gives us a rough idea of what constitutes a near-death experience, but if we are going to scientifically examine them, we need a more systematic approach. The work of Dr. Bruce Greyson, at the University of Virginia, is helpful in this regard. Greyson has spent many years studying the psychiatric aspects of these experiences. In one study, he analyzed

Table 1: This is roughly the sequence of what often happens during near-death experience.

Beginning	Recognizing the crisis
	Feelings of peace
	A noise (buzzing)
	Dark tunnel
	Light
to	Out of the body
	Meeting others
	Being of light
	Life review
	Reaching a border
End	Returning

the accounts of sixty-seven people who had experienced a "close brush with death." He took eighty characteristics that he found in their accounts and from them formulated sixteen questions that fell into four categories: Cognitive (thoughts), Affective (feelings), Paranormal, and Transcendental. Each question was assigned a value from zero to two for a possible high score of thirty-two. A total score of seven was the minimum necessary to consider an event a near-death experience. This questionnaire has turned out to be a reliable scientific tool. (See the "References and Resources" section at the end of the book.)

My investigative team at the University of Kentucky used Greyson's scale on our fifty-five research subjects (see Table 2), who scored from seven to twenty-eight. No one scored the maximum of thirty-two. The average score was about sixteen. Different characteristics combined in each individual's experience, and there was no single aspect of near-death experience that was considered absolutely necessary for eligibility in our study except that the subject felt that he or she had faced a life-threatening crisis.

The common features our subjects experienced were peace, unity with the universe, brilliant light, vivid sensations, leaving their bodies, entering another world, or reaching a border and returning to this

Table 2: The nature of near-death experience in our 55 research subjects.

GREYSON NEAR-DEATH EXPERIENCE ELEMENT		
Category	**Topic**	**Percent of subjects**
Cognitive	Time sped up	62
	Rapid thoughts	44
	Life review	36
	Profound understanding	60
Affective	Felt peace	87*
	Felt joy	64
	Felt harmony or unity with the world	67*
	Saw or felt brilliant light	78*
Paranormal	Vivid sensations	76*
	ESP-like awareness	31
	Saw future scenes	29
	Separated from body	80*
Transcendental	Entered another world	75*
	Encountered mystical being or presence	55
	Encountered deceased or religious spirits	47
	Reached a border or point of no return	67*

* Denotes that at least two-thirds of subjects experienced that feature.

one. Greyson applied his scale to cardiac arrest survivors and found fewer paranormal experiences, such as visions of the future, than our sample. On the Greyson scale Patrick scored only four, three short of being a scientific near-death experience. I would score Seymour's near-death experience eight. Here I think we see a shortcoming of Greyson's tool, since in my estimation Patrick, like Seymour, had a spiritual experience that would meet the approval of William James.

Our research group was intrigued with how infrequently the paranormal appeared in these experiences. Extrasensory perception or viewing scenes from the future were the least common features: only 31 percent of our group had some ESP-like awareness; only 29 percent saw the future. In Greyson's study, these phenomena occurred 10 percent or less of the time.

As we collected accounts in our study, it was clear that each one was shaped by life experience, cultural background, and individual and shared biology. The following cases illustrate this.

The Origins of Hell

Oliver Sacks referred Margaret, an Australian woman, to me. Margaret e-mailed me about her near-death experience. She had had a brain aneurysm as a young woman. To clearly see the aneurysm in preparation for possible surgery, she had an angiogram. Dye was injected into her brain arteries so they could be visualized by X-ray.

This usually routine procedure was excruciatingly painful and dangerous in Margaret's case because, for medical reasons, it was done without sedation or anesthesia. Each time dye was injected into her, Margaret was overwhelmed by pain and her heart stopped. This happened thirteen times by her count.

Margaret wrote that with each injection she found herself slipping feet first down a long tunnel: "Even though I had a husband and two small sons that I cared about, they didn't matter to me as much as my desire to go down round the corner at the bottom of the tunnel which was so very alluring. I knew once I got around that enticing warm corner everything would be fine."

Margaret described the tunnel's walls as soft and silky, shining with pink light. The farther down she went, the more the light reddened, until it became a rich ruby color that emanated from around a corner at the tunnel's bottom. Margaret described this light as the "most remarkable thing" about the experience and associated it with an "Edwardian brothel with red velvet curtains, a piano, and an open fire, [and] of course, tea and crumpets! There was the welcoming buzz of people socializing, and I knew I would fit right in."

Past life regression? Perhaps.

Margaret said she was sure that what she was experiencing was death. Over the years, she noted, a common response when she describes this experience has been "Well, you know where you're going, don't you?"

The red light at the bottom of the tunnel "fits so nicely with our pervasive culture of a red hell downwards," Margaret wrote. "Some people react seriously strangely if I mention it [and] have even gone so far as to completely avoid me because they think I am damned, while I am myself just amused at the weirdness of it all, and very, very curious."

It didn't surprise me when Margaret reported that she was subject to unusual experiences in the transition between sleeping and waking. "I have always had trouble telling if I am awake or asleep when I first wake up," she wrote. "We have a news service come on as our wakeup call, and I frequently get my current dream totally mixed up with whatever has been happening in the world. My husband thinks it funny, but I find it very disconcerting."

Like so many of the people I have read about and interviewed who have had an near-death experience, Margaret said her experience was "realer than real."

"You'll be pleased to hear I have no plans to found a religion based on my experience," she wrote. "But I can see how this could be tempting."

The Skeptic

You might think that near-death experiences with spiritual content can only strike those who are already spiritually predisposed. Not so.

The headline in London's *Sunday Telegraph* read: "What I Saw When I Was Dead . . ." A shocking statement from anyone, but particularly from Sir Alfred Ayer. Ayer was born into a wealthy family, educated at the finest English schools, and spent most of his professional career teaching at the University College London and Oxford. He wrote on William James and produced a biography of Britain's foremost contemporary philosopher, Bertrand Russell. Both were famous atheists.

Ayer wrote in the *Telegraph* about the impact on his thinking of his near-death experience. In 1988, at the age of seventy-seven, hospitalized with pneumonia and intolerant of the institution's food, he dined on provisions brought in by family and friends. This proved to be his near fatal mistake. He carelessly threw a piece of smoked salmon down his throat the wrong way and began to choke. His heartbeat plummeted. Then his heart stopped. The attending doctor later told Ayer that he had "died in this sense for four minutes." The medical staff was able to resuscitate him. His condition stabilized, but he spent many days in a coma.

Although the medical details are murky, it is highly unlikely that all blood flow ceased to Ayer's brain during the four minutes his heart stopped; his brain did not die, otherwise he would not have been able to make such a remarkable recovery.

In true Ayer fashion, the first thing out of his mouth when he emerged from his coma was: "You are all mad!"

In the newspaper account Ayer described how he had tried to cross the River Styx, succeeding only after a second attempt. "I have not wholly put my classical education behind me," he wrote. "It was most extraordinary. My thoughts became persons. The only memory that I have of an experience, closely encompassing my death, is very vivid. I was confronted by a red light, exceedingly bright, and also very painful even when I turned away from it. I was aware that this light was responsible for the government of the universe. Among its ministers were two creatures who had been put in charge of space." It seems these ministers had failed in their duties "with the result that space, like a badly fitting jigsaw puzzle, was slightly out of joint . . . It was up to me to put things right. I also had the motive of finding a way to extinguish the painful light. I assumed that it was signaling that space was awry and that it would switch itself off when order was restored."

An obstacle confronted Ayer: the ministers were nowhere to be found. And even if he came across them, how would he communicate? Pondering this conundrum, he recalled Einstein's general theory of relativity. He reasoned that he should "treat space-time as a single whole. Accordingly, I thought that I could cure space by operating on time. I then hit upon the expedient of walking up and down, waving my watch, in the hope of drawing their attention not to my watch itself but to the time which it measured. This elicited no response. I became more and more desperate, until the experience suddenly came to an end."

Such was Ayer's experience, which I score twelve on Greyson's scale. Ayer correctly surmised that "my brain continued to function although my heart had stopped." Consequently, he found no evidence that consciousness continues after death but reflected on a phenomenon that he had studied most of his career: how we carry our self-identities through the successive physical bodies that we occupy, from birth to death. Nearly all of our cells (except neurons) are replaced every seven years. And even the molecules are exchanged that make up all the cells that remain with us throughout our lives.

Ayer took the opportunity to point out a prevalent fallacy that every near-death experience, including his own, seems to spawn: life after death proves that God exists. He argued that since this life is not proof of God, why should the next be different? There might be evidence of God in the next life, but, he reasoned, "we have no right to presume on such evidence, when we have not had the relevant experiences." (Ayer didn't say what those experiences might be.) Although his experience left a strong impression on him, he found it only "slightly weakened my conviction that my genuine death, which is due fairly soon, will be the end of me, though I continue to hope that it will be." He added that his experience had "not weakened my conviction that there is no God."

Later, Ayer would take pains to correct what he meant by "slightly weakened." His belief that there is no life after death was not weakened, but rather his inflexible attitude toward that belief was; for the first time he felt it might be worth examining. He also reexamined his ideas about memory, resurrection, and reincarnation. He remained an atheist to the end and maintained that even if consciousness continued after our brains died, we would bear witness to the "triumph of dualism," the separation of brain and soul, yet "we should still have no good reason to regard ourselves as spiritual substances."

Ayer was reading Stephen Hawking's *A Brief History of Time* in the hospital just before his near-death experience. Perhaps this explains why Einstein's theory of relativity and space-time were so central to his experience. Ayer's story shares the narrative qualities of the one Paula told about emerging from the coma after her head injury; it is less like the abrupt experiences of Seymour and Patrick. These differences could reflect different sets of neurological circumstances, but it's difficult to determine what's happening in the brain from first-person accounts. A major obstacle is illustrated by Ayer's episode. There is no way to know precisely when Ayer had his experience, no marker to tell us when it occurred. It could have happened while he was choking or subsequently, when he was comatose. *When* exactly near-death experiences happen is critical to our investigation.

Ayer was convinced that his near-death experience came about because his brain continued to function during cardiac arrest. It is interesting that Ayer's is one of few cases I know where a boundary was reached, crossed, and then re-crossed. Still, I don't see a neurological reason why Ayer crossed his River Styx when others haven't crossed their border. Jesus, on the other hand, stopped Satan from abducting Joe. My guess is that the differences in the two experiences are cultural rather than neurological.

The Great Psychiatrist

Another of the twentieth century's great thinkers also had a near-death experience, although he was not an atheist like Ayer. Carl Jung, the famous psychiatrist and author, thought our brains are naturally religious and believed in the power of intuition. In his foreword to *The I Ching or Book of Changes*, Jung cast aside the Western scientific notion of causality as intellectually narrow. He used the oracle throughout much of his life for purposes of divination and to explore his subconscious.

In his autobiography, *Memories, Dreams, Reflections*, Jung wrote about his near-death experience, which occurred when he was sixty-eight years old, after a heart attack, when, as he tells it, he was unconscious.

"I would never have imagined that any such experience was possible," he wrote. "It was not a product of imagination. The visions and experiences were utterly real; there was nothing subjective about them; they all had a quality of absolute objectivity."

Jung found himself in space, surveying the earth below, which was "bathed in a gloriously blue light." He had life review—the sense of reviewing everything he had ever experienced. He saw oceans, continents, snow-covered mountains, and a temple floating in space. He approached the temple with the certain knowledge that he would encounter a room filled with people he knew and meet others who would tell him the meaning of his life. Then his journey took an unexpected turn. His physician, Dr. H., "in primal form," floated up from Europe, where Jung's physical body lay, to deliver a message. There had been a protest against Jung leaving earth—he must return. "The moment I heard that," wrote Jung, "the vision ceased."

Jung was profoundly disappointed to find himself back on earth and in his body. Yet the experience transformed him. He wrote: "After the illness a fruitful period of work began for me. A good many of my principal works were written only then . . . Something else, too, came to me from my illness." He found it easier to accept "the conditions of existence as I see them and understand them."

After his near-death experience, relations between Jung and his physician became strained. Jung both resented Dr. H. for bringing him back to life and at the same time he worried about a terrible premonition he had had. Jung thought that because Dr. H. had appeared in primal form during the experience he was going to die soon. Not only that, but, thought Jung, he "would have to die in my stead."

Such was, indeed, the case. "In actual fact, I was his last patient," wrote Jung. It seems that on the day Jung finally got up after his prolonged bed rest, in another part of town "Dr. H. took to his bed and did not leave it again."

Jung, who viewed dreams as portals into the Universal Unconscious, spent his recovery having "glorious" and "enchanting" nightly visions based on traditions and rituals he knew from his past, only to find himself waking to a drab world each morning. The visionary power of his dreams slowly faded, only to reappear after his wife died. She came to him in her prime, wearing a dress made by Jung's cousin, a spiritual medium. "I knew it was not she, but a portrait she had made or commissioned for me," he wrote.

Although Jung was a deeply spiritual person, his near-death

experience did not constitute evidence that going beyond near-death experience would bring him into an afterlife. "We lack concrete proof that anything of us is preserved for eternity," he wrote. "At most we can say that there is some probability that something of our psyche continues beyond physical death. Whether what continues to exist is conscious of itself, we do not know." As for the apparitions that had come to him during his near-death experience, he wrote: "the question remains whether the ghost or the voice is identical with the dead person or is a psychic projection, and whether the things said really derive from the deceased or from knowledge which may be present in the unconscious."

One of the salient features of Jung's experience is that, unlike Patrick's and Seymour's, it relied heavily on his past memories. Also, unlike Patrick's and Seymour's but more in line with Ayer's crossing the River Styx, Jung's near-death experience was full of intricate imagery that was woven from his past. It shows us that any brain mechanism that explains near-death experience must be capable of weaving autobiographical memory into a rich narrative.

Near-Death Experiences in Children

The role of autobiographical memory in the near-death experiences of children is very different from its role in adults' NDEs. Children simply do not have the accumulation of life experience that adults do, and they don't experience the kind of life review, or their lives flashing before their eyes, that is a fairly common feature in adult accounts. There are other differences as well—their vision of the afterworld includes castles and rainbows, often populated with pets, wizards, guardian angels, and like adults they see relatives, and religious figures, too.

Beyond the difference in the specifics of their accounts, the very interesting question remains: why would children, many who have little to no understanding of death, have near-death experiences in the first place?

In his book *Closer to the Light*, pediatrician Melvin Morse analyzed this kind of experience in twenty-six children. Many core features of near-death experience in adults were found: a sense of being dead, seeing a light, and out-of-body experiences. Several children also reported encounters with dead relatives, living teachers, family members, angels, and godlike beings. They felt peace and joy. Nearly half of Morse's group made a conscious decision to return to life. One child reported seeing Jesus (who looked like Santa Claus).

Their near-death experience had a powerful impact on some of the children. One told Morse: "I am not afraid to die anymore because now I know a little about it. You have to tell all the old people about this." A ten-year-old girl took it on herself to counsel children who were dying of leukemia after her experience. In a number of instances children assuaged their fear of their own death and their parents' anxiety.

A panoramic life review was absent. Also absent was an alteration of time and a feeling of unity with the universe. That absence of unity may be more apparent than real since the unity experience is ineffable and defies description by even the most articulate adults.

If near-death experiences are enabled by a particular neural system, then it is reasonable to expect to find that system in children's brains as well, although it may not yet have achieved full adult development. But we should not expect that a child's description of the experience will have the same narrative qualities as an adult's. A child's point of view and breadth of experience is very different.

Near-Death Experiences as Symbolic Narrative

One characteristic particularly impressed me in our fifty-five research subjects' accounts of their near-death experiences. At its very root and center the near-death experience is a *story.* This happened and then that happened. It is often a story with a beginning, middle, and end. Most often its narrative is about a journey and a return, an old story that, as mentioned above, stretches back at least as far as the ancient civilizations in Mesopotamia, Egypt, and Greece, where corpses were ritualistically prepared for their journey into the afterlife.

Carol Zaleski, a professor of religion and a scholar of near-death experience, has rigorously drawn the historical backdrop to our modern narrative of returning from near death. She found that today's near-death experiences are more like medieval accounts of returning from the dead than like the nineteenth-century séances that William James attended. She treats these experiences as a product of a religious (or spiritual) imagination that symbolically communicates meaning. This view, of course, is not endorsed by people who hold near-death experiences to be true in the most literal sense.

People who glimpse the afterlife sometimes return with the message that virtue has its rewards. Plato presents this theme in *The Republic*, recounting the myth of Er, a solider slain in battle who returns to life on his funeral pyre and tells how he left his body and traveled to the place where souls are judged. The judges tell him to return to earth with a message for mankind to lead just and pious lives full of good deeds. The reward will be heaven.

Plato, of course, lived in an age when symbolic narratives were taken very seriously. Not so for Jung who, in his chapter "On Life After Death," written after his near-death experience, lamented: "The mythic side of man is given short shrift nowadays." Jung argued that myths (particularly about life after death) are critical connections between the subconscious and conscious minds: "We are dependent for our myth of life after death upon the meager hints of dreams and similar spontaneous revelations from the unconscious." He said that if we are cut off from mythic imagination, the raw material for a probing intellect, then "the mind falls prey to doctrinaire rigidities."

And though he realized that some experiences make us contemplate life after death in a new way, he also warned us against the seductive power of these stories, which can make us prone to confuse the symbolic with the real.

When I envision a cup of coffee in my mind, it can have all the essential characteristics of smell, color, taste, and warmth, but I would never expect my mind's image to hold hot coffee. Perhaps the things seen in near-death experience have a kindred symbolic reality.

Witnessing the Truth

So far I have taken near-death stories at face value, without questioning whether they have been accurately conveyed to us. Christopher French, a London psychologist, justifiably asks if false memories form an important part of the experience. Memory can be wildly selective and distorting. French points to the mountain of scientific research demonstrating that eyewitness testimony is unreliable, particularly when it involves extraordinary events. This point has been graphically demonstrated in recent years by the Innocence Project, an international organization that uses DNA evidence to exonerate people who have been wrongfully convicted of a crime. Through the organization's efforts, hundreds of innocent people have been released from prison, usually after having been convicted of murder or sexual assault based on solid eyewitness testimony.

Witnesses don't necessarily lie, but they can make colossal mistakes. Jennifer Thompson-Cannino was one such case. On a summer night in 1984, a man broke into her apartment and sexually assaulted her. She got a long, close look at her assailant, vowed to remember his face, and subsequently identified him in a police lineup. Yet DNA evidence produced years later proved beyond a doubt that she had identified and sent the wrong man to prison.

Most of us regard ourselves as expert witnesses to our own experiences, and spiritual experiences are no exception, especially because they're often so vivid and compelling.

Still, in my practice as a clinical neurologist, and in the medical literature, it is stunning how often our perception can be distorted without giving us the slightest hint that this has happened. We've already seen how we can be our own false witness to the most basic perceptions of self. Many examples have been given of the whimsically dictatorial power of our brains over what we think of as real. Here is one more example: a stroke can suddenly render a wife unrecognizable to her husband of fifty years.

This is to say that the physical brain is perfectly capable of producing completely credible encounters with the transcendent. I'm not saying that we shouldn't reach beyond science for explanations of the mystical or visionary. But we have to distinguish when we're operating in an area of faith or speculation rather than empirical evidence or proof.

Neurologists have examined how memory structures are damaged when blood flow to the brain is interrupted, a common trait of near-death experiences. Memory is the first brain function injured and the last to recover after cardiac arrest; the damage can be temporary or permanent, mild or severe. And even under the best circumstances, the brain systems that construct our visual experiences and the memories that come from them start with a falsehood. You are not actually "seeing" Mona Lisa when you look at the painting. The light reflecting off the canvas gets only as far as the retina on the back of your eyeball. The retinal image is upside down. The eye and brain convert the image to nerve impulses that are transmitted to the occipital lobe, where they are fabricated into a mental image; turning the Dora Maar of brain activity into the Mona Lisa of experience.

From a neurologist's vantage point, the science of our visual system could be said to resemble what we see in the cave of Plato's famous parable. In the cave, shackled prisoners could see things only as shadows cast on the wall, and they were blind to the world at their backs. In our world we see light waves reflecting off objects, not objects themselves. Similarly, in Plato's cave, we don't see actual objects, only their shadows on the cave wall. This is all we really can perceive of the things around us, testimony to the fragile way the brain processes what we see, retains fragments, and then retrieves experience and tricks us into assuming that our subjective experience is objective reality.

When it comes to extrasensory perception or prescient visions (like Jung's premonition that his physician was going to die in his stead), I share William James's openness to the unseen universe and fully embrace the idea that extraordinary things are possible. Extraordinary claims, however, require extraordinary evidence. Susan Blackmore—a psychologist who has herself had out-of-body experiences—closely scrutinized extraordinary near-death experience claims.

Among the most startling of these were near-death cases where the blind saw what was going on around them. However, what they saw could have been explained by lucky guesses or subtle suggestibility. Jung could have easily overheard others speak of his physician's failing health. We just don't know. Information can be absorbed during these experiences from eyes half open or conversations overheard while medical personnel and others are preoccupied. Patients may appear dead while their brains are very much alive.

Although people can and do lie about just about anything, my experience has been that the vast majority of near-death experiences are faithfully rendered. I have no intention of picking apart near-death or any other spiritual experience story by story and bit by bit. I take them as they are given to me, but with eyes wide open to the ways our brain distorts through subtly filtering, shaping, and interpreting the raw data of our experiences.

Stories Can Take Us Only So Far

It's only recently, in Western culture at least, that near-death experiences have been recognized as relatively common and acknowledged as a type of spiritual experience. This is reflected in what seems to be the inexhaustible media interest. It was, in part, the pervasiveness of near-death experiences through history and cultures that predisposed me as a neurologist to look for the cause in the fundamental brain processes we all share, in the instinctual way the brain reacts to crisis and the threat of death.

We all wonder what's going to happen to us after we die, and stories of traveling to that threshold and returning, meeting dead relatives or archetypes, and encountering supernatural light are innately compelling, all the more so because they are so vivid and dramatic.

The characteristics of near-death experiences measured by the Greyson scale—changes in thinking and mood, visions and unearthly light, unity with the universe, out-of-body experiences, life review—combine to tell us that wide expanses of the brain are engaged during these experiences. What systems work in tandem and can tie these regions together?

As compelling as I find each near-death experience narrative, these stories can only take us so far into the brain. Let's see how much further we can probe with all the tools and knowledge neuroscience has on hand right now. We are learning more about how the brain works every day.

And this is true for the neurobiology of spiritual experience as well.

5

THE BRAIN AT DEATH'S PORTAL

LIGHT AND BLOOD

"If an explanation should be found, then near-death literature would lose its power to elicit a sense of wonder."

—Carol Zaleski

"It does not do harm to the mystery to know a little more about it."

—Richard Feynman, theoretical physicist, Nobel Laureate

Although near-death experiences are common, we really don't know how often the events described occur: after all, if the experience goes too far, then dead men tell no tales. Studies have shown that 6 to 12 percent of cardiac arrest survivors who made a good language and memory recovery in the hospital report having had a near-death experience. In these studies, researchers found no underlying conditions that trigger near-death experiences—no particular medication, chemical imbalance in the blood, or resuscitation pattern. Nor do these experiences seem to be linked to the length of time it took to revive a patient or how long he or she was unconscious (which reflects how long the brain was deprived of blood and oxygen). Most research indicates that the patient's age, personality, and gender don't matter either. There is a suggestion that near-death experiences are more likely in people under sixty or people who have had more rather than less oxygen in their blood, but this greater tendency is slight. If true, it could simply reflect that better memories can be formed by younger brains, as well as brains that suffer less damage because they have more oxygen and blood.

The theory has been proposed that perhaps endorphins, our brain's natural narcotics, are released during medical crisis and cause these experiences. That would explain the euphoric mood that frequently accompanies the experience. But it does not explain why near-death experiences are also accompanied by out-of-body, brilliant light, or why they often assume a narrative character.

Another theory about what causes these experiences has to do with the neurochemical system N-methyl-D-aspartate (NMDA), a chemical used by nerves to communicate with one another, which excites the brain. The effects of the drug ketamine, sometimes used in anesthesia and acting through the NMDA system, mimic some features of the near-death experience, including hallucinations and, on rare occasions, out-of-body experiences. But the ketamine experience, in and of itself, is a far cry from the narrative accounts of near-death experiences. Rarely is there an out-of-body experience or euphoria. Perhaps most important, patients given ketamine do not experience "the light" that is so often a core feature of the experiences we have been considering. Even if the NMDA system comes into play during a near-death experience, we are still left wondering why it's triggered in the brain.

The most neurologically bizarre explanation I have heard came from the Pulitzer Prize–winning astronomer Carl Sagan, who thought that near-death experiences were released memories of birth, which are deeply stored within all of us. Reliving a birth experience that occurred before the brain was physically capable of forming the synapses necessary for memory strikes this neurologist as wonderfully poetic, but woefully unscientific.

In a poll of nearly two thousand people interested in near-death experience, 45 percent said they believed they were encounters with a "non-earthly" spiritual world; 20 percent attributed the experiences to body and brain function; and the rest felt that the cause remained elusive or they registered no clear opinion. These numbers are only approximate, but they do illustrate a divide in America today in how the experience is perceived.

Blood Flow and the Brain

Before coming up with my REM intrusion hypothesis as a way to explain what was happening during near-death experiences, I believed near death must engage more than one brain region or system. Its manifestations were just so varied.

One of the first questions my study asked was what physical events preceded the near-death experience. In my fifty-five-subject study group, fainting was at the top of the list, followed closely by heart disturbances, near drowning, violent trauma, and events surrounding surgery.

The overwhelming majority of conditions that my subjects shared had the potential to cause a temporary interruption in blood flow or oxygen to the brain. This is significant for a number of reasons.

Normally, 20 percent of the blood the heart pumps sustains the

Table 3: The circumstances prompting near-death experience in 55 research subjects.

Cause	Number
Blackout/Faint (or nearly so)	10
Heart disturbance	8
Near drowning	8
Car accident	8
Physical head trauma	5
Events surrounding surgery	5
Stroke	3
Fall	2
Severely low calcium	2
Carbon monoxide poisoning	1
Drug overdose	1
Latex Allergy	1
Lightning	1
Total	55

brain. If the blood flow is reduced to a third of its normal supply, the brain remains immediately active, but after ten to twenty seconds, it loses consciousness. The brain sustains no injury, even if this flow rate lasts for hours. At these marginal flows, a person may slip in and out of consciousness. I have witnessed this many times in patients and in research subjects who are losing blood flow to the brain but remaining conscious for more than ten seconds.

Permanent brain damage occurs thirty minutes after blood flow is reduced by 90 percent or more. If there is zero blood flow to the brain for a minute or two, the coma that follows may persist for hours or longer. After four or more minutes without blood flow, temporary brain disturbance becomes permanent injury. In short, the brain tightly regulates the blood flow it receives because, second to second, brain function and life depends on it.

In the aftermath of the severest form of fainting, cardiac arrest, the heart may be successfully resuscitated, but we end up losing a great deal of the brain. The limbic system memory structures, including the hippocampus, are most vulnerable to lasting damage. After the hippocampus, other regions of the limbic system take big hits, including the cingulate gyrus, which has a profound impact on survivors' mental and emotional state, often leaving them not only with severe memory loss but apathy as well.

Let's look at the most common cause of near-death experience—fainting. If we understand the causes of the near-death experience, we will begin to be able to see how the experience itself manifests in the brain.

Fainting and ESP

My first encounter with what could lead to a near-death experience began as a teenager's trick.

On a boring Sunday afternoon I was hanging out with friends in the basement of one my buddies' houses. Someone suggested we try to make ourselves pass out—an idea that appealed to our adolescent sensibilities. My friends tried and failed, but I suspected I could do better. I squatted down and hyperventilated until my lips tingled and my vision dimmed. Then I sprang up and pressed my lips against my forearm, bearing down as if I were trying to clear my ears on an airplane.

This maneuver was beautifully designed to deprive my brain of blood. I fainted, and one of my friends caught me as I fell.

When blood returned to my brain, it was as if I was emerging from a deep sleep. I didn't know where I was or what had happened. Then I became excited. As I blacked out, a vision had come to me. My father was at the open door of our house, angrily summoning me home. This was confusing since my father rarely rounded up us kids. That was my mother's duty.

Although the faint had not been frightening or unpleasant, I had the fear of my father in me and beat it hastily home. Late Sunday afternoons were normally not a time for family outings. But sure enough, as I approached the house, my father intercepted me, angrily reminding me that the family had plans and I had made everyone late.

There was no way that I could have physically heard or seen my father during my faint. But in some extraordinary way, I thought I had actually seen him calling for me. Later I came to realize that something had triggered the memory of our family's plans during the faint, although the break in routine had completely slipped my mind until I had my vision.

Before you try such a trick, I hasten to add that the Centers for Disease Control and Prevention has recently warned about a more dangerous version of fainting. The choking game, or dream game as it is sometimes called, is practiced by mostly adolescent boys who use ligatures around their necks to strangle blood from the brain long enough to bring on a dreamlike euphoria. Choking the brain of blood is also sometimes a part of sexual practices. To get the stranglehold just right is tricky, and when it goes wrong the adventure can end in death.

Fainting in the Laboratory

I discovered many years later that my fainting experience was not that unusual. Dr. Thomas Lempert and his colleagues in Berlin have studied fainting, recording experiences that were very much like mine.

And they were also among the first researchers to connect fainting with near-death experience.

In the early 1990s, Lempert prompted forty-two young, healthy subjects to safely faint in his laboratory, meticulously recording the event on high-speed video. He carefully debriefed the subjects afterward. What he found surprised everyone. Sixty percent of the fainters had visual hallucinations, from a simple haze to colored patches and bright lights. Some saw realistic scenes that involved familiar places, situations, and people. The visions seemed to occur while the subjects were unconscious or in the borderland between consciousness and unconsciousness.

The hallucinations were also sometimes aural. About a third of Lempert's subjects heard sounds, such as rushing, roaring (a typical aspect of near-death experiences), screaming, and human voices.

Four subjects, nearly one in ten, had an out-of-body experience.

Fainting was a "neutral" or "positive" emotional experience for 83 percent of Lempert's subjects. Only 7 percent found it distressing. Often the fainters described feelings of weightlessness, detachment, and peace. Some compared it to drug or meditation experiences. Two subjects said fainting reminded them of their near-death experiences. One remarked about his faint: "I thought that if I had to die in the very moment I would willingly agree."

Lempert's team compared the experience of their subjects to Moody's descriptions of the near-death experience. Surprisingly they found *no real difference* between the two types of experience. To these investigators, fainting in the laboratory and a near-death experience in crisis looked about the same.

The implications of Lempert's study are enormous. Fainting is quite common. Upward of 100 million Americans have fainted at least once in their lives and therefore could have had an experience that, in many respects, was identical to a near-death experience.

Table 4: Table is based on the study of Dr. T. Lempert and his colleagues, who compared his results with fainting with the near-death experience narratives compiled by Moody.

Experience	Near-death (percent)	Fainting (percent)
Out-of-body	26	16
Visual perceptions	23	40
Audible noise or voices	17	60
Feeling of peace and painlessness	32	35
Appearance of light	14	17
Life review	32	0
Entering another world	32	47
Encountering preternatural beings	23	20
Tunnel experience	9	8
Knowledge of the future	6	0

Why should these two states be so neurologically similar? What common physiologic process underlies them? Lempert and his team thought a "dying" limbic system sets them off.

Physically the limbic system straddles the brainstem deep within the cortical mantle of the cerebral cortex. The limbic system is where emotions (foremost fear) and their bodily expression, instincts, and memories all come together. Paul Broca is credited with the idea of a limbic lobe, named because structures form a limbus or border deep within both sides of the brain. Even now, neuroscientists can't fully agree as to what brain regions make up the limbic system. When examined under the microscope, the cortex devoted to the limbic system is primitive compared to the more newly evolved cortex that is responsible for our "higher" cerebral functions like reasoning. The

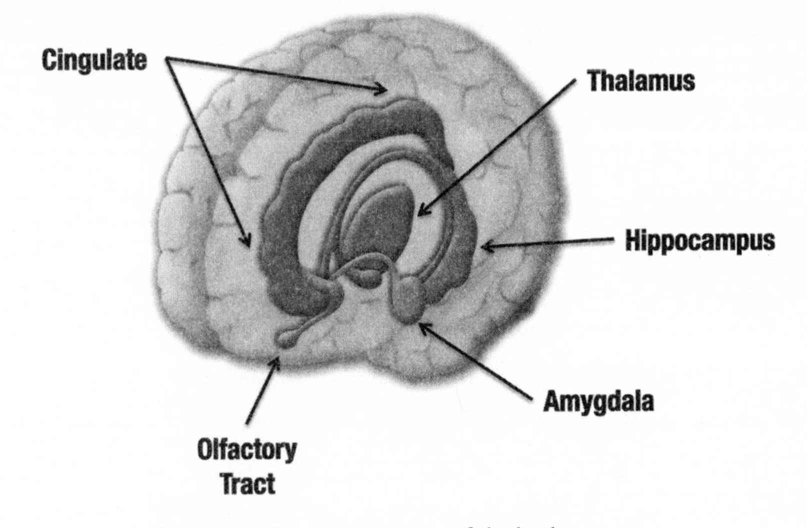

Figure 9: Major components of the limbic system

limbic cortex has fewer than the six layers of nerves found in the advanced cortex. Instead of a six-lane freeway, information traffic travels over four lanes in the limbic system.

Since limbic structures are scattered throughout the brain, the limbic system is best conceived not in physical terms but by looking at its emotional and memory functions.

The limbic brain area underlies nearly all aspects of our emotional lives, including pleasure (e.g., food and sex) and motivation. It governs our internal organs and provides our visceral responses to emotion such as a fast heart rate and sweating. This role brings the limbic system to the fore when we faint or our heart arrests. Of all our senses, smell is the one most closely integrated with the limbic system since it has been so important to the survival of species that preceded us. The limbic system is at the front line of our instinctual survival reactions, linking it tightly with the arousal system to keep us conscious at the right time.

But this leaves the question of *why* our limbic system behaves in such a way when it's deprived of blood. Why do so many people see a brilliant light during NDEs? Light has little to do with the limbic structures. How about tunnels? Why is it that mostly dead people appear in an NDE? Clearly, there is more going on in NDEs than a limbic system starved for blood.

In a later project, Lempert's research team monitored the eyes in fourteen people who fainted. The eyes of every person remained *open* as they went from consciousness to unconsciousness and back again to consciousness. Fainters—or anyone with low blood flow to the brain, for that matter—may register more of what is going on around them than we have realized. I am reminded of Dr. Moody's Mrs. Martin, who overheard her radiologist and was alert but unable to respond. And Jan, who was awake and alert during her surgery after her gunshot wound. When we study patients with minimal consciousness, we find brains that are far more active than we would expect. It's clear that people who faint or have cardiac arrest take in far more than physicians suspect.

A great deal of what happens in the brain during near-death experiences comes about because of a reaction to the crisis of having low blood flow, regardless of how briefly. But surprisingly, at least to a neurologist, not everything about the near-death experience is produced by the brain.

The Tunnel Mystery Solved

What could be the physiological basis of the tunnel that appears so often in NDEs? Evidence to answer this question comes from a human centrifuge that was housed in the basement of the Medical Sciences Building at the Mayo Clinic in Rochester, Minnesota, during World War II.

Leading up to the war, the speed and maneuverability of fighter planes dramatically improved, so much so that the plane's effectiveness in combat became severely limited by the force a pilot could withstand before blacking out during a tight bank or pulling out of a steep dive.

The pressure pilots feel is measured in multiples of the force of earth's gravity (g). A man who weighs 150 pounds on the ground weighs 300 pounds in a 2g turn and can barely stand up. At 3 g's or more, if precautions are not taken, blood drains from the skin and the face becomes "pale as a ghost" before blackout and loss of consciousness occur.

Blackout and unconsciousness were serious problems for our pilots in the war. The military called on Dr. Edward Lambert, an esteemed neurophysiologist, for help. Lambert was soft-spoken, gentle, and supersmart (when I spoke with him, I was reminded of Charles Sherrington whose synapses bridge the gap between neurons). He wanted to understand how a pilot's brain reacted to g-force. Because it was not practical to take these measurements in a plane, he devised a cockpit that was constructed and placed on a massive centrifuge that simulated the g-force of flight. The centrifuge reached several g's in two to three seconds and usually maintained that force for fifteen seconds. It turned out that a lot happened in those fifteen seconds.

Lambert and his colleague Dr. Earl Wood gauged a pilot's ability to see lights placed across his visual span as he whirled around. During the first three seconds, when exposed to a force designed to bring on fainting, the pilot's peripheral vision began to dim, and it was completely lost by the fifth second. What is particularly important for our discussion is that the pilot could only see within a small circle in front of him—it was as if he were looking through a tunnel! At eight seconds, his central vision was lost, and he was completely blind but conscious, producing the "blackout" that comes right before fainting. The blindness was temporary, and his vision was fully restored as soon as the g-force abated. By fine-tuning the force, Lambert could keep the pilot in the "tunnel" or make him blind for as long as the doctor wanted, without causing unconsciousness.

If the forces were increased slightly during blackout, by half a g, hearing and consciousness were lost. The pilot became "limp" and later "frequently recalled dreaming" during or after blackout, although it was impossible to determine the exact point that happened.

Lambert wondered what caused the tunnel and the blackout—blood draining from the eyes or the brain? He used an elegantly simple technique to find out, applying suction to the inside of special goggles to lower the pressure inside them and permit blood to flow more freely to the eye when the heart could barely pump hard enough to keep blood flowing to the head. When a pilot wore these goggles, his vision did not black out on his way to losing consciousness. Instead, he went directly from consciousness to fainting. Mystery solved. When not enough blood is pumped to the head, the eyes fail first, causing tunnel vision before the brain fails and unconsciousness occurs.Neurologists are aware of what is referred to as "tunnel vision" in routine clinical practice. Not only do many patients have tunnel vision as they are about to faint, they can also have tunnel vision when they hyperventilate from anxiety, causing the eye's blood vessels to constrict and choke off blood supply. The eye's retina is exquisitely sensitive to poor blood flow, even more sensitive than the brain when blood flow ceases to the head. When this happens, peripheral vision fails. We have all experienced this when we stand up too quickly. Blood pools in the legs, and for a moment the heart can't pump enough of it to the head. Our vision grays and we become light-headed.

Besides standing up too quickly or whirling in a centrifuge, a great many things can cause fainting, including fright and pain. If an injury, like Jan's gunshot wound, causes severe blood loss, the body may not have enough blood to pump to the brain, which can also result in a faint. If enough blood is lost, shock and death will follow.

If blood pressure and with it blood flow falls sufficiently during fainting from any cause, then as the face becomes ashen, with blood draining from the skin, the eye also becomes pale. Vision fails in the periphery first, creating a tunnel, before failing completely into blindness and unconsciousness. Depending on how quickly and to what extent blood flow is reestablished, the process can stop at any stage, with the person having prolonged tunnel vision or blindness, or being forced directly into unconsciousness. Often, during this process, memory, sensitive to blood loss, is disrupted. Only fragments of what induced the faint are recalled. Or there is no memory of the actual faint at all.

After fainting, heart disturbances accounted for the second greatest number of near-death experiences in our study sample. Initially, when blood flow stops, the brain can't tell the difference between harmless fainting and full-blown cardiac arrest. In the first ten seconds or so of each event, the eyes and brain go through the identical process. From the brain's perspective, the major difference between fainting and cardiac arrest is the extent to which blood flow to the brain is disrupted. In cardiac arrest, the flow is more likely to be closer to zero and remain low for a longer period of time than during a faint. How soon and how much blood flow resumes determine how much of the brain is damaged. This damage can range from none to brain death.

It's important that we now examine exactly what constitutes "brain death," because there is massive confusion on this point, especially where near-death experiences are concerned.

Not Death

The brain does not die during a near-death. Being *near-death* is very different from *returning* from death. Although this is assumed in the scientific literature, some authors imply that physical death is possible while consciousness is retained. There is no empirical evidence yet for this extraordinary claim.

A most improbable source has led to confusion on this point.

In 2001, the prestigious journal the *Lancet* published an article by the cardiologist Pim van Lommel and his colleagues on their near-death experience research, which, as it turned out, quickly became and *remains* one of the most important scientific publications on the topic. Van Lommel interviewed 344 consecutive patients resuscitated from cardiac arrest; it turned out that 62 of them had had a near-death experience. "All patients had been clinically dead," they wrote, "which we established mainly by electrocardiogram records." As a neurologist, I was stunned. As we have seen, after blood flow stops, the brain goes along quite nicely for ten seconds or so. It is not dead. Only after that ten-second point does the brain begin to malfunction, but it still doesn't approach death for several minutes, even when it has zero blood flow.

The brain is nowhere near physically dead during near-death experiences. It is *alive* and *conscious.*

Brain death happens by the death of cells. When lack of blood kills a brain cell, calcium rushes in, causing the cell to rupture like a water balloon being burst by a pin. Once the cell is ruptured, there is no putting it back together. New brain cells do not grow to make up for those that have died. The brain, as a whole organ, dies cell by cell, and when a critical number of the 100 billion brain cells reach a point where they have ruptured, and the few that might remain can't sustain life, the brain is dead. Van Lommel's near-death experience patients were not "clinically dead"; they were in a state that resembled the faint of Lambert's pilots and Lempert's subjects.

A great deal has been made of the fact that we sometimes don't know the exact moment that the brain dies. After all, the thinking goes, if we don't know when the brain is actually dead, how can we be sure that near-death experiences don't demonstrate evidence of life after death?

This question is far less dramatic when we understand how the brain works.

A single cell dying does not tip the entire brain toward death, and with massive brain cell death, a handful of living cells scattered here and there will not sustain the individual's life. At the same time, huge numbers of brain cells in the thalamus and cortex can be lost, leaving a person permanently comatose, yet still with a beating heart and the ability to breathe.

This is precisely what happened to Terri Schiavo. If the brainstem alone dies, then cells in the cortex are unable to bring about consciousness because they have lost the arousal system. This is what makes Dr. Nicholas Schiff's study so exciting—implanted electrodes, it seems, can take over some brainstem functions (see page 55).

It is true that the precise magical moment of brain death may elude us, or it may not even exist, or it may not be important. There is a gradual continuum between one neuron dying and the point where every cell in the brain has ruptured. As cells die one by one, the borderland that a brain can enter and then viably return from is large and murky.

Dr. Sanjay Gupta chronicled how finding the boundaries of this borderland can get *really* murky. In the mountains of Norway, orthopedic resident Anna Bagenholm, an experienced backcountry skier, was on an outing when she fell into a stream and was trapped under the ice for ninety minutes. Her blue and lifeless body was pulled from the icy water and transported by helicopter to a hospital more than an hour away. Her body temperature was fifty-eight degrees, and it was hours before her heart began to beat again on its own. Yet, eventually, she was revived so successfully that she now works as a radiologist. Anna's brain was never dead, although it looked like it when she arrived at the hospital. Instead, the freezing water had put her neurons in suspended animation, and kept them from bursting like balloons.

Mind and Body Revisited

If mind and brain can exist separately, then, as Ayer reminds us, we have gone a long way to proving there could be an afterlife. This is what seems to be happening during out-of-body experiences. Our consciousness feels as though it has somehow detached itself from our physical self. But, I'm disappointed to say, this is an illusion.

Throughout history, out-of-body experiences have influenced us, finding expression in folklore, mythology, spirituality, literature, art, and religion. The experience of floating above your body and other even more extravagant flights from the body have been presented in both ancient and contemporary cultures as proof of an immaterial spirit that is, in some sense, you. One of the people most strongly identified with dividing mind from body or mind from brain is the seventeenth-century philosopher René Descartes, inventor of calculus and perhaps philosophy's most famous phrase: "*cogito ergo sum*" (I think therefore I am). Descartes drew sharp distinctions between our "animal" bodies and our "human" mind. He considered animals conscious but soulless automata, bundles of unthinking, purely reflexive responses that were incapable of language, thought, and wisdom, which were attributes of the eternal spirit and immaterial rational soul.

Descartes identified the small pineal gland deep within the brain as the "seat" of the soul, which brought the spiritual mind to the material brain. He regarded this roughly spherical pea-size gland hanging delicately down into the brain's fluid-filled cavities, between the right and left hemispheres, to be the brain's center.

We may not know all the reasons why Descartes chose the pineal gland, but it is the only brain structure where the left and right sides of the brain are fused. For Descartes, this unity was important. He reasoned that only a singular brain structure could mediate nonfragmented or unitary consciousness. Because it linked the brain's two halves, the pineal gland could reign over the left and right sides of both the brain and the body. Descartes thought the pineal, like other glands, distributed "animal spirits" within the brain's cavities, swirling in currents and eddies that caused muscles to move the limbs, eyes, and face. My day-to-day clinical practice, where I witness self-fragmenting from brain disorders, makes it hard to accept Descartes's division between mind and brain. Belief in experience beyond the brain is based on faith and not science. Yet Descartes was prophetic when he wrote: "Whatever I have up till now accepted as most true I have acquired either from the senses or through the senses. But from time to time I have found that the senses deceive, and it is prudent never to trust completely those who have deceived us even once."

Truer words were never written when it comes to out-of-body experiences.

Self and Brain Part Ways During Brain Surgery

Wilder Penfield was one of Charles Sherrington's most promising students at Oxford. When Penfield returned to his native Canada, he devoted his neurosurgical career to exploring the human cerebral cortex by stimulating it in hundreds of completely awake patients to see what happened.

Penfield did his experiments in the operating room. Depending on where Penfield put his probe, his subjects experienced multicolored lights and simple shapes; full-fledged fear or sounds of passing automobiles. His probe sometimes evoked familiar music or memories. One subject reported: "[The] dream is starting—there are a lot of people in the living room—one of them is my mother." Another patient exclaimed: "I feel queer, as though I were floating away." After Penfield moved the probe, she said: "I have a queer sensation as if I am not here."

Penfield used William James to contextualize his discoveries. He likened his subjects' life experience to James's "stream of thought," or, as Penfield put it, "a river forever flowing forever changing." Penfield believed the stream of consciousness could be activated "as though it were a strip of cinematographic film recording the sight and sound, the movement and meaning which belonged to each successive period of time."

Penfield thought his stimulation probe replayed the strip, a pathway formed by physical changes in nerve bundles during the original experiences. The physical changes that he thought were so important have turned out to be in Sherrington's synapses. His stimulating electrodes set off experiences that continued as long as he supplied the electrical current.

In 1955, Penfield published one of his most interesting cases, which has a direct bearing on what we know about near-death experience today. Penfield described "V," a thirty-three-year old man afflicted with frequent seizures that originated in his right temporal lobe. These were no ordinary seizures. When they struck, V felt a whirling sensation (vertigo), as though he were on a merry-go-round. He also had a strong sense of déjà vu and was seized by feelings of fear that came with tightness in his stomach and rectum.

One of the most unusual features to V's attacks, which caused him great anxiety, was that they were brought on by smells. Air currents wafted perfume molecules into V's nose, where they were greeted by receptors and nerve endings. The nerve impulses traveled to the temporal lobe, where brain pathways brought the odors to consciousness.

In V's case, these pathways not only brought smells to consciousness but ignited his seizures as well. On two occasions, visits to a department store perfume counter caused him to seize. The fragrance of a woman sitting next to him on a bus or passing in a crowd posed a real hazard, so V took measures against any chance encounter with stray odors. Unsavory smells, alas, cast no spell on him.

Convinced that V's seizures originated from a brain region that he could surgically remove, Penfield readied V for the operating room. During surgery, most of V's temporal lobe was exposed to the doctor's probe, which was inserted nearly an inch deep into the temporal lobe where it begins to merge with the parietal lobe. Immediately, V told

Penfield that he had a "bittersweet taste" on his tongue. V became confused, making smacking and swallowing motions. Penfield cut the current, but it was too late: he had triggered a seizure.

"Oh God! I am leaving my body," V exclaimed.

When his abnormal brain rhythm shortly stopped, V seemed himself again.

Penfield asked V if this experience resembled his usual seizures.

"A bit, sir," V replied. "I had the fear feeling."

Penfield moved the probe slightly and again inserted it deeply. This time the stimulation caused a whirling vertigo sensation. After stimulating the area again, V felt as if he were "standing up" (he was actually supine on the operating table).

These two regions, it seemed, contributed to V's seizures. The first gave him his fear sensation and the second produced vertigo, a sure sign that a seizure was on its way. The areas responsible for vertigo, fear, out-of-body experience, déjà vu, and V's seizures all went the way of Penfield's scalpel.

Physical stimulation of the physical brain caused V to have out-of-body sensations. Neurologists were intrigued, but we would have to wait fifty more years for a fuller explanation of what happened to V in Penfield's operating room.

Out-of-Body in the Laboratory

Out-of-body experiences are extraordinarily common in the general population. In a survey of more than thirteen thousand Europeans, 5.8 percent reported they had had this kind of experience. At least one out of every twenty people has had one, or, conservatively, 15 million Americans. Susan Blackmore, an English psychologist who has studied the phenomenon, reviewed several smaller surveys and has concluded that they probably occur even more often.

Few physicians have appreciated this fact. I was recently at a dinner party for officers in our national neuromuscular society. During the conversation I mentioned that out-of-body experiences are far more common than most people realize. To illustrate my point, I said to the person next to me that the chances are that someone in the room had had an out-of-body experience.

The president of the society overheard my remark.

"I've had one!" he exclaimed.

The room fell silent. He said that as a young boy of six he was lying in bed and briefly viewed his body from above. He didn't recall having been particularly disturbed by the experience, although its memory was now faint. He mentioned that during that period in his life, when he woke he was often completely paralyzed for a few terrifying moments.

An out-of-body experience can be defined as a disembodied sensation, seeing the world from a perspective different from the body's actual location. Autoscopy is a special form of out-of-body experience where one's body is on view, often from a higher vantage. Rarely, if ever, is the body viewed from below. These are fleeting experiences, reflecting an unstable brain state. When they occur spontaneously, such as during the transition between wakefulness and sleep, or in near-death experiences, their duration is commonly seconds, in some instances minutes. They rarely if ever last for hours or days. They happen infrequently, usually only once or twice in a lifetime, although some of my research subjects had out-of-body experiences nearly every week.

They often happen in the borderland between wakefulness and sleep. They can be brought on by a lack of sleep and danger. They can occur with or without a near-death experience, while awake during surgery, during recovery from surgery, when fainting, during seizures, with migraine headaches, while flying jet aircraft, during high-altitude mountain climbing, and under hypnosis. Nearly all out-of-body experiences happen when a person is lying down. This has been my experience with patients and research subjects and is borne out by other researchers. This is an important clue about the physiology of these experiences.

Paul Firth and Hayrunnisa Bolay wrote about an unusual case. A healthy twenty-eight-year-old physician had an out-of-body experience while descending from a high-altitude trek. At first he had a powerful sensation that another person was nearby. The sensation was so powerful that he turned repeatedly to see who was following him and began to converse with his "companion." Soon, "While walking, he felt as if his legs were moving of their own accord and his torso was elongated. He felt detached from his person, as if he were observing himself from a distance. He was aware of the hallucinatory nature of his experiences. These anomalies lasted approximately ten minutes before disappearing spontaneously."

This physician was upright and walking when he left his physical body.

In the category of out-of-body experience called heautoscopy, a double appears and the person having the experience may not even be able to decide whether he or she is in the double's body or his or her own.

It may come as a surprise that most of these experiences have no spiritual significance for the individuals who have them. The explaining left hemisphere does not always interpret out-of-body experiences in a spiritual way. The mind leaving the body is sometimes just weird, not godly.

Neurologists have discovered that out-of-body perspectives are created by disrupting how the brain puts sensations together to form the self's body schema (which is not, I should note to avoid confusion, the map of sensations found by Penfield that is changed with phantom limbs). We do not know which of the many sensations are necessary to create the body's schema, but sensing where the body is (where your right foot is now, for example) and body movement and touch are important. And we are not talking about just a single limb or body part—out-of-body experiences are about a displacement or misapprehension of the location of the whole body.

In out-of-body experiences our vision must be affected, as well as the sense of where we are in the earth's gravitational field, which has to do with the vestibular organs in the middle ear. Vestibular sensation gives us our sense of motion and balance and allows us to remain steady as we stand on a rocking boat. When the vestibular system goes awry, we experience a whirling dizziness, vertigo, and seasickness.

To get a sense of where our body is, and hence where our consciousness resides, our brain has to bring together multiple bodily sensations. The brain integrates vision, messages from our inner ear, as well as feelings from Penfield's map: the position of our arms and legs.

When we look at patients like V, we have our first clue where these sensations get disconnected. Evidence points to the area where temporal and parietal lobes meet: the temporoparietal junction.

The temporoparietal area is a bit of a no-man's-land. When you put your hand on the side of your head above your ear, your hand is close. The area is situated above the temporal lobe, which is concerned with sound and vestibular sensation. It is below the parietal lobe, which processes our limb sensations of touch and position, and in front of the visual brain, in the occipital cortex.

Olaf Blanke and his colleagues in Switzerland, building on Penfield's work, made an unexpected discovery while mapping a patient's brain in preparation for surgery. A hundred or so electrodes were placed on top of the brain in a sheetlike grid. At each site, a physician applied a tiny electrical current, and the response it caused was used to "map" brain functions, indicating to the surgeon what part of the brain could be safely removed (operations of this sort are usually done to control seizures).

One of Blanke's patients was a forty-three-year-old woman who suffered from incapacitating seizures that originated in her right temporal lobe. To find out where in the lobe her seizures originated, a large grid of electrodes was temporarily placed over her right brain. When physicians stimulated the brain point by point, she unexpectedly had strong vestibular feelings of "sinking" or "falling."

As the current was increased, she exclaimed, "I see myself lying in bed, from above, but I only see my legs and lower trunk."

The research team turned off the current and she immediately returned to her body. When the current was turned back on, she instantly felt "lightness" and "floated" six feet or more above her body. Current on and she immediately left her body; current off, she returned.

Current applied elsewhere prompted her left arm and both legs to shorten and, after she bent her knees slightly, caused her legs to feel as if they were moving rapidly toward her face.

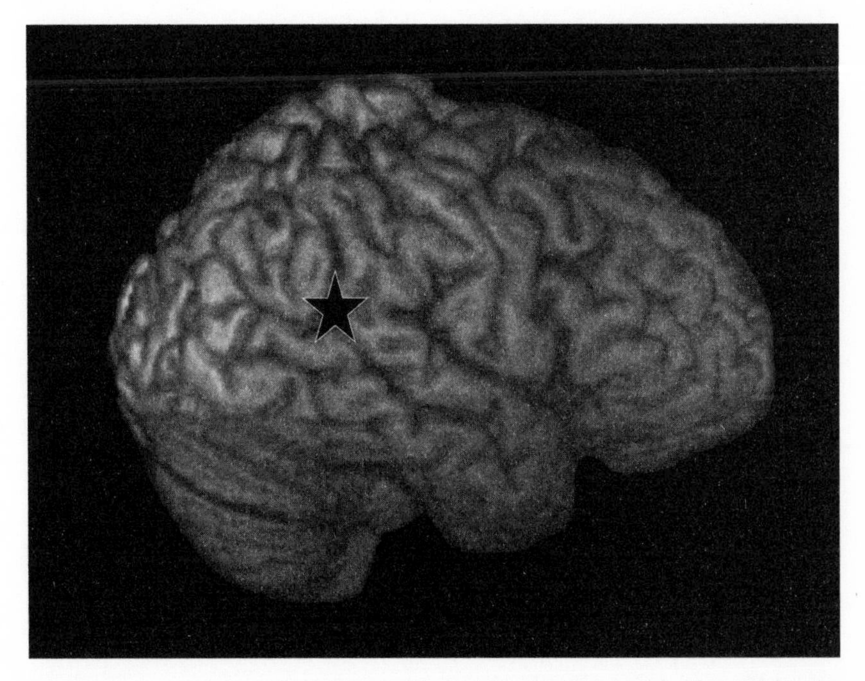

Figure 10: Stimulating the brain in the temporoparietal region (denoted by the star) reliably causes an out-of-body experience.

The Out-of-Body "Up" Button

While the current is on, her brain is not able to bring the sensations of where she is in space together in the temporoparietal region. Although the stimulating electrode is supplying current, it may not be activating cortical circuits. Instead, it may be shutting down circuits, inhibiting them from integrating the necessary sensory information that places conscious perception within her body.

The woman's sense of being in or out of her body came and went with the mechanical predictability of turning on a light switch. The person manning the switch moved her consciousness at will, meeting the bedrock criteria of scientific verification—being able to predict and control results. It was as if the elevator up button for an out-of-body experience had been discovered.

Given this very clear laboratory result, we have even less reason to jump to the unnecessary conclusion that consciousness leaves the physical body during a near-death experience. Rather, consciousness has lost its bearings in relation to bodily position, touch, gravity, and motion. It is no neurological accident that the cortex specializing in vestibular sensation is close to the cortical region provoking out-of-body sensations. Neurologists have known for years that vestibular symptoms accompany out-of-body experiences, and now have a better idea of why.

But there is an interesting leap to make here: when the integration of our bodily sensations becomes fragmented, our consciousness finds itself anchored in vision, our most dominant sense. Using our vision or visual memory, consciousness temporally projects itself onto our visual map, what we see around us, when the sensation map of the rest of our body is obscured.

Out-of-body experiences may be a connection with the cosmic consciousness—but we do not need to rely on that explanation. Penfield and Blanke's electrodes did not create spiritual epiphanies. What they discovered was that it takes just a small trickle of electrical current in the right place to induce such an experience.

Other factors, such as a temporary lack of blood or oxygen to the brain, may also interfere with sensory integration in the temporoparietal region and cause out-of-body experiences. Blanke's subjects, as well as many of the cases of near-death experiences with out-of-body experiences that we've discussed, were able to take in events going on around them. They simply lost their sense of where their bodies were in space.

The temporoparietal region is also important for taking the first-person perspective, perceiving ourselves to be the agent of our actions. This region may also help us appreciate other people's minds, and is important to the pathways that allow us to empathize with others. An emotion that is no stranger to spiritual experiences. Many cortical areas are critical to put the self together, like the prefrontal and posterior cingulate, which we discussed earlier in relation to self-reflection. We can add the temporoparietal junction to the list of regions critical to the conscious self and how we perceive the location of that self in relationship to our physical body.

Chasing a Spiritual Illusion

Why, you might ask, if the evidence in neuroscience is overwhelming, did a research team of cardiologists in the ER rooms in multiple medical centers suspend cards with writing or symbols that could only be read if someone's consciousness had indeed left their body and was actually hovering near the ceiling?

They hoped, of course, to prove against the evidence, that consciousness does, indeed, leave the body during a near-death experience.

Their first efforts have been extensively reported in the media. Not knowing the devil in the all-important details of how the study will be conducted limits what I can say at this point. However, it strikes me as unlikely that proof of an afterlife will be revealed through a card trick. Consciousness outside the brain, though an ordinary claim of faith, remains an extraordinary and unnecessary claim in neuroscience.

No Brain Activity?

Van Lommel and his team not only confused "clinical death" with syncope (loss of consciousness due to insufficient blood flow to the brain) and reversible brain injury, they took this confusion one big step further, citing the experience of Pam Reynolds as extraordinary evidence that a near-death experience can happen "during the period of flat EEG" (which measures brain waves), when there is no electrical brain activity—in other words, "brain death."

Even without these extraordinary claims, Pam's story is astounding. When she was thirty-five, a major artery at the base of her brain had formed a huge ballooning aneurysm. It would have caused a catastrophic stroke if this artery had ruptured (the aneurysm's size alone could have damaged the brain even if it hadn't ruptured).

The aneurysm had to go, but it was so large that it required the highly dangerous technique of draining all the blood from Pam's brain so the surgeons could operate on the artery. Only by putting Pam's brain in a type of suspended animation could she survive the operation. To do this, her doctors turned off her brain metabolism by lowering her brain temperature to sixty degrees and giving her high dosages of barbiturate sedatives. If metabolism can be slowed so much that the brain doesn't require oxygen and glucose, then it can remain alive without blood for a long time.

Pam remembered being brought into the operating room early on the morning of the operation. She was soon under anesthesia and prepared for surgery. Pam's cerebral cortex and brainstem electrical activity were closely monitored. The neurosurgeon swept the scalpel in a large half circle so the scalp over nearly the entire right half of her head could be peeled back, revealing the skull. To reach the brain and blood vessels, a skull flap had to be removed with a bone saw, which is a handheld device, pneumatically powered like a jackhammer. As the saw cut through her skull, Pam awoke, and came to view her surgery as if she were sitting on the neurosurgeon's shoulder. First she heard what she said was the sound of a musical note being played. It felt as though the sound was pulling her out of the top her head. She recalls looking down at herself from above in the operating room, and a feeling of acute awareness, "the most aware I have ever felt in my life." Then she found herself looking at the proceeding as if she were sitting on her surgeon's shoulder. Her vision was bright and clearer than "normal vision." She saw them turn the saw back on and she heard the whirring sound it made.

After he finished sawing, the neurosurgeon removed the bone flap, exposing Pam's brain and its coverings. While this was taking place, other surgeons struggled to find the large arteries in Pam's groin so her blood could be drained. Pam said she was aware that the doctors were having trouble finding her veins and arteries. She said she felt as though she was being pulled upward in a vortex, spinning around. She noted that it was "like a tunnel but it wasn't a tunnel." At that point, she heard her grandmother calling to her. She didn't so much hear it, "it was a clearer hearing than with my ears," she says. Her grandmother beckoned to her and she went toward her down a "dark shaft" at the end of which a light burned brighter and brighter. At the end of the shaft she was enveloped in light and she met other people, presences also bathed in light. She recognized her grandmother. She said she wanted to be absorbed by the light but her overriding concern was to return to her life. She had children whom she knew needed her.

Pam was fed "something sparkly" by her deceased relatives; then her uncle took her back to the end of the tunnel, where she saw something terrible. Her body "looked like a train wreck. It looked like what it was: dead. I believe it was covered. It scared me and I didn't want to look at it. It was communicated to me that it [returning] was like 'jumping into a swimming pool.'" Pam's uncle "pushed" her back toward her body. "It was like diving into a pool of ice water," she said.

The neurosurgeon made his way around Pam's brain to the aneurysm. He could work on it, but first Pam's brain had to be put in suspended animation. Her body temperature was lowered to sixty degrees and barbiturates were given to cause all brain activity to cease. Her heart then stopped and she was placed on a heart-lung bypass machine. At this time, both her cerebral cortex and her brainstem were electrically silent. After the surgical team had drained all the blood from her head, the aneurysm deflated like a balloon, allowing the neurosurgeon to clip it shut at its base.

Pam awoke once again in the operating room, but this time after the operation was over, while the surgical wounds were being closed. "When I came back they [the surgeons] were playing 'Hotel California' in the background," she recalls.

Pam's saga was written about by cardiologist Michael Sabom, who had the opportunity to review her medical records. Sabom said the weight of scientific evidence indicated that during surgery Pam's soul had separated from her physical form and transcended the material world.

Scientifically, this is an unwarranted assumption. There is no doubt in my mind that Pam awoke during surgery in the same way that Jan did. Much of her experience can be explained by simply being awake and not fully anesthetized. Being awake during surgery is, fortunately, rare, occurring in approximately 0.18 percent of patients. I know that patients have near-death experiences as they come out of anesthesia.

To me, the most intriguing thing about Pam's experience is that she had her experience as she was regaining consciousness during surgery. Why would her temporoparietal region be selectively turned off like one of Dr. Blanke's patients in his laboratory? Knowing what we do, there is no need to evoke a supernatural explanation for what Pam saw and heard. She no doubt registered overheard conversations and these combined with memories of being brought into the operating room before the operation. The strongest support for the argument that her consciousness actually left her body was her close, but flawed, description of the saw used to open her skull. But she'd had an opportunity to visualize the saw and other instruments when she was wheeled into the operating room before the surgery began, and after the surgery, as she talked about her experience, information was subtly conveyed to her and incorporated into her memory.

How would I explain her vivid experience while she had a "flat EEG"? Here, too, a more worldly explanation is not only likely but obvious. Pam's near-death experience occurred *before* her brain was put in suspended animation. There is no reason to believe she met her grandmother or had any other part of the experience when her EEG was flat. After reviving her brain and the anesthetic wearing off, Pam remembers waking to "Hotel California."

As I've already mentioned, during fainting and cardiac arrest, patients are taking in much more than doctors realize. They overhear conversations when they seem "dead," and their impaired brains form memories. Pam's was a real, profoundly important spiritual experience. But it falls far short of scientific proof that consciousness transcends the material world.

If I have one piece of advice for physicians it is this: patients who appear to be unconscious may be much more aware of their surroundings than anyone suspects. I have taught this lesson to my team time and time again as we make the rounds in the intensive care unit.

A Shadowy Presence

One of the features of many spiritual experiences, especially in those near death, is sensing the presence of another person or supernatural being. Paula (from Chapter 1) sensed and recognized her grandfather who had passed away. Both Ayer and Jung encountered spiritual beings. Pam met her grandmother.

My friend Jake suddenly awoke at 3 A.M. one night, felt a breath on his face, smelled his mother, and strongly sensed her presence. At that apparent moment she died a continent away. I will later explain more fully how this could happen in terms of the brain. For now it is enough to know that sensing another being is a brain function.

Using brain-mapping techniques, Blanke and his colleagues discovered shadowy presences lurking in the temporoparietal junction of a twenty-two-year-old woman who suffered from intractable seizures. Her brain was being mapped to prepare her for surgery. When a small current stimulated the temporoparietal region, she did not have an out-of-body experience. Instead, a person appeared behind her. She described this person as still, silent, young, and shadowy. "He [was] almost at my body, but I [did] not feel [him]," she reported. With the next stimulation, a man sat behind her, clasping her unpleasantly in his arms. This stimulation was repeated a number of times from two different sites. Each time the presence was in a different position, although it was always behind her.

During one of these shadowy encounters, the "person" tried to take a card from her hands that she was reading. "He wants to take the card," she said. "He doesn't want me to read [it]."

The "person" was always in nearly the same position as she was, leading the investigators to conclude that she had created a ghostlike double.

Sensed presence is similar to an out-of-body response and can also be caused by disrupting the brain in the temporoparietal junction. In both cases, the current disrupts the many sensations that are necessary to put the physical and conscious self together. Disruption of the temporoparietal junction by a lack of oxygen could be why the physician hiking at high altitude in thin oxygen sensed a person behind him before his out-of-body experience.

It is not a great leap from sensed presence to supernatural encounters during a spiritual experience or a crisis such as being near-death. But temporoparietal disruption may not account for all the presences that can be sensed. A fuller explanation of brain function during such experiences might rely on memory and the narrative abilities of other brain regions.

In all these instances what we see is that the self is a synthetic process that pulls different components distributed throughout the brain into the illusion of a unified whole. The neurologist in the hospital or in the laboratory often sees firsthand how fragmentary this illusion is, and how vulnerable it is to certain kinds of disruptions. We see this again and again.

Frightened to Near-Death

In the next chapter we will see that people can, literally, be frightened to death. So it only seems natural that they can be frightened to near-death as well. Some investigators have asked whether someone actually needs to be in true medical danger to have a near-death experience. Having a spiritual experience as one is falling from a height results from fright and not near-death as I've been discussing it—at least not until one hits the ground. Facing possible death at gunpoint, before the gun is fired, is another example of a "fear-death" experience. The person is not medically threatened at the moment, and his or her brain has not suffered impairment from such things as low blood flow during cardiac arrest. But there may be no real difference between the fear-death and the near-death experience.

Justine E. Owens and colleagues, psychiatrists at the University of Virginia, investigated this question. They examined the medical records of fifty-eight people who had had near-death experiences. Twenty-eight of those had had a real medical crisis, while the other thirty were not medically endangered at the time. Surprisingly, the results showed the experience in both groups to be almost identical. In both, people went through a tunnel and had similar thoughts and emotions. Sixty-eight percent had an out-of-body experience, and it did not matter if they were medically near-death or not. Turns out you can be scared out of your skin.

It is interesting that the only difference in the experience of the two groups was that those people who had experienced a real medical crisis during their NDE were much more likely to see unusual light.

I draw two conclusions from this study. The first concerns the light. Near-death experience physiology must have an intimate relationship with light. Secondly, the circumstances leading up to these experiences are critical. Life experiences, combined with what happens to us during fainting or cardiac arrest, explain a great deal about what's happening in the brain at this time. Yet we need to go further and look at life-threatening circumstances that act on the brain as it approaches death's doorway. After all, fear of death goes back a long way in evolutionary time. It has molded our lives and brains and spiritual experiences since the beginning.

6

THE ANCIENT METRONOME

THE TEMPO FROM FEAR TO SPIRITUAL BLISS

"Death is indeed a fearful piece of brutality."

—Carl G. Jung

The brain is our most glorious organ. To survey the majesty of all human accomplishments is to survey the brain's majesty. Beethoven's symphonies, Shakespeare's plays, Plato's philosophy, Einstein's scientific insights—the brain dazzles us with its capacity and power.

With all its magnificence, it's easy to forget that the brain's primary role is to keep us alive each and every second. It regulates each breath we inspire.

Hundreds of millions of years ago, when all creatures on earth were simple and primitive, nerves were arranged like a net throughout the body. This allowed symmetrical animals like the jellyfish to propel themselves about the sea. Later, when mouths first appeared, large numbers of these nerves concentrated up and down the body to control not only movement and the gut, but also the hearts and gills for the respiration that developed afterward. The nerves in the body regulating these new organs brought their neurochemistry with them, and they eventually grew in number and complexity, forming spinal cords and brainstems to manage the survival of fish and amphibians; procuring food and escaping danger and procreating. It is the chemistry of the body's nerves, nerves controlling gut, heart, and lung, that laid the foundation for the chemistry of our brains. Much later, when mammals first appeared, they used a brainstem that was eerily similar to the vertebrates that preceded them.

Early mammals were small creatures and prey to the larger and far more numerous reptiles. The human brain proved to be supremely capable at this task of survival. What other species can live above the Artic Circle, in equatorial jungles, on remote ocean islands, and in parched deserts? Survival in these conditions relies more on the intelligence from our cerebral cortex than on visceral reflexes in the brainstem.

In the last chapter, we looked at aspects of near-death experience that can be explained by low blood flow to the head. Now we turn to the brain's arousal system in the primitive brainstem that controls our conscious states. Our attention will also focus on the limbic system, which controls our reactions to danger and is tightly linked to the arousal system.

The arousal system is not only key to our survival but has a vital role in near-death experience and, perhaps, other types of spiritual experience as well. We shall see that it makes sense that the spiritual doorway in the brain is primal and connected at its root to the origin of life, and it is fitting that our most enduring transcendental moments engage parts of our brains that we share with mammals, reptiles, and birds.

Consciousness and Survival

The brains of mammals have grown increasingly complex, but the primitive brainstem, an intricate confluence of nerves, is remarkably unchanged.

We see that same brainstem 2 million years ago in *Homo erectus*, the first of the human species that was very much like us. *Erectus* lived in closely knit hunter-gatherer bands. Babies were born early and immature, probably because their large skulls and brains wouldn't otherwise fit through the birth canal. Social bonds were forged that supported a long, nurturing childhood. With his long legs, *erectus* was the first hominid to run swiftly, and unlike the apes before him, he sweated instead of panting to cool his body. He was probably the first hominid to use fire. He also ate meat, certainly scavenging but probably also hunting it. And *erectus* was hunted: he was a dietary staple of many large predators.

Let's follow a little band of *Homo erectus* for a day in their life on the plains of East Africa 1.5 million years ago, to begin to unravel how the ancient survival mechanisms in our brains are tightly connected to our experiences of transcendence, and particularly to NDE.

> Dusk is settling over the boundless savannah. Beside a streambed, sheltered by a stand of tall trees, rests a troop of *Homo erectus*. There are, perhaps, fifteen members of this small group, including children.
>
> It is the end of a good day. Earlier the six women and four children dug for tubers they had discovered the day before. After digging for the tubers, they found a grove of fruit trees and then returned to camp before the sun had scorched the plains.
>
> The five men had been busy as well. They left early in the morning, setting out with a systematic plan across the plains. In the early afternoon, several miles from camp, they spotted an old antelope straggling behind the herd, worn out after it had barely outrun a pack of wild dogs. The hunters coordinated their downwind ambush. Crouched over, stalking with utmost care, they slowly and silently surrounded the lone, exhausted beast. When they were in position, the strongest of them stood upright and hurled a heavy stone that found its mark, stunning

the antelope long enough for the other men to rush in, pounce on it, and deliver killing blows to its head.

The hunters barely paused to savor the moment. The last thing they wanted to do was attract the attention of other predators. One of the hunters quickly threw the antelope over his back and the men loped off toward camp. They moved athletically, trading the carcass from hunter to hunter. The clan was jubilant when they returned. A few sharp blows to a stone cobble deftly knocked off sharp flakes that were used to butcher their kill.

The troop has been in this spot for four days. They would soon move camp to follow the season's last fruit harvest. But now it was a time to eat and rest. The day ended in calm satiety.

Darkness falls and the troop settles in for the night. A loud cry freezes everyone. From the parameter of the clearing in which they are camped, one of the men sees a menacing shadow in the brush twenty yards ahead. *Erectus* knows its outline well—*Homotherium*, a lion-size saber-toothed cat. Normally they see the big cat in prides on the open plains, running down elephants. They know this solitary hunter was probably drawn to the camp by the scent of the fresh antelope kill.

The men's eyes scan the brush, tracking the cat's every movement, and looking for other, hidden cats. They don't have much time. If the cat strikes, it will be on them in seconds. Their eyes search for stones and the heavy sticks the women used to dig tubers from the soil. Instantly, they know the troop's safety lies with the large overhanging ledge above their camp. From that vantage, by working together, they can mount a strong defense against the predator. If the troop is forced to make a stand in the open, it will mean almost certain death for one of them.

Everyone remembers through lore or experience to suppress the urge to run: a running target triggers the cat's chase-and-attack instinct. The women and children move deliberately up the rock to safety first while the men gather below, standing guard with sticks raised and rocks in hand. One by one, each man makes his way upward.

The cat has her own ideas. Seeing the hominids' defensive posture, she sprints for the antelope carcass, clamps it in her jaws, and bounds silently away.

Wide Awake

What happened in *erectus*'s brain during this incident? And what does it tell us about spiritual experience?

In the moment of crisis when the *Homotherium* approached, the brain had to be in the proper conscious state. If you'll recall, from a neurological viewpoint, the brain has only three states from which to choose: awake, REM sleep, or non-REM sleep. It's generally not a good survival strategy to fall asleep or begin dreaming in the face of danger. It is the brain's arousal system that is responsible for keeping us awake at these moments.

Being awake is such an obvious requirement for survival that we don't even think about it, and it seems silly to be talking about it. Yet the brain can't take wakefulness for granted. For millions of years its circuits have guaranteed that the lion's roar is a wake-up call. The brain that doesn't wake up becomes someone's meal—evolution selects against it.

A part of the brain important to regulating our consciousness is a small cluster of nerve cells called the locus coeruleus, on each side of the brainstem. It is the source of nearly all the brain's nor-Adrenaline, which affects the brain very much like its almost identical cousin, adrenaline, affects the body. The locus coeruleus is a tiny oblong clump of sixteen thousand cells two millimeters wide by fifteen millimeters long. These cells are darkly pigmented. They stand out to the naked eye, and they send microscopic projections to nearly every recess in the brain. A single cell's tentacles (axons) branch to widely scattered targets.

The locus coeruleus helps interrupt the brain from its routine and readies it for something new and important, a trait that is important when we confront a dangerous world. One of the locus coeruleus's particularly strong connections is to the limbic system's amygdala and hippocampus. When the lion's silhouette triggered our ancestors' adrenaline surge, the locus coeruleus readied the hippocampus to retrieve and form memories essential to their survival. This is a direct link between the adrenal surge and our autobiographical memories.

Through its connections, the locus coeruleus plays crucial roles in arousal and conscious states, paying attention and responding to stress.

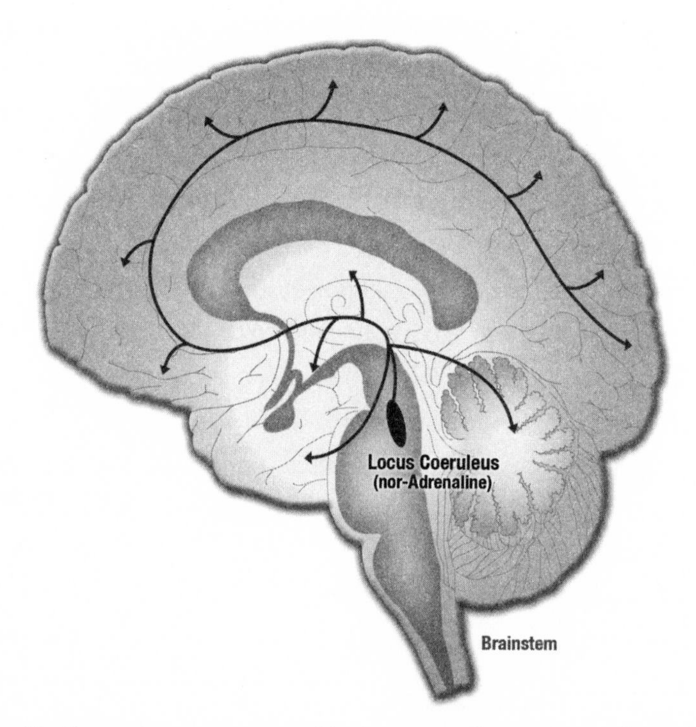

Figure 11: The ancient metronome. The locus coeruleus is a tiny cluster of neurons that distributes nor-Adrenaline to nearly every region of the brain through a microscopic web.

Allan Hobson, a Harvard professor, psychiatrist, and one of the world's foremost experts on dreams, helped discover how the locus coeruleus shifted activity to and fro as we moved between wakefulness and REM sleep. Today Hobson describes the locus coeruleus as the brain's metronome, firing in a set rhythm during every waking moment. It slowly ticks away if we are quiet and inattentive (stumbling around in the morning after a long night or sitting through a boring speech). If something arouses our attention, it accelerates. It fires rapidly when our bodies or minds are stressed, bathing the entire brain in nor-Adrenaline. When we are doing something that requires alertness and focused attention, like putting together a puzzle or intently watching a TV program or reading this book, it fires at a moderate rate.

Remarkably, locus coeruleus firing is not only tightly linked to behavior—it actually *anticipates* what we are about to do. That means that whatever the locus coeruleus is doing, it must be absolutely essential to the brain's behavior. That is not to say that the locus coeruleus *causes* the behavior, but it is critical to making the behavior happen. These thirty-two thousand neurons set a powerful stage for the other 100 billion in the brain.

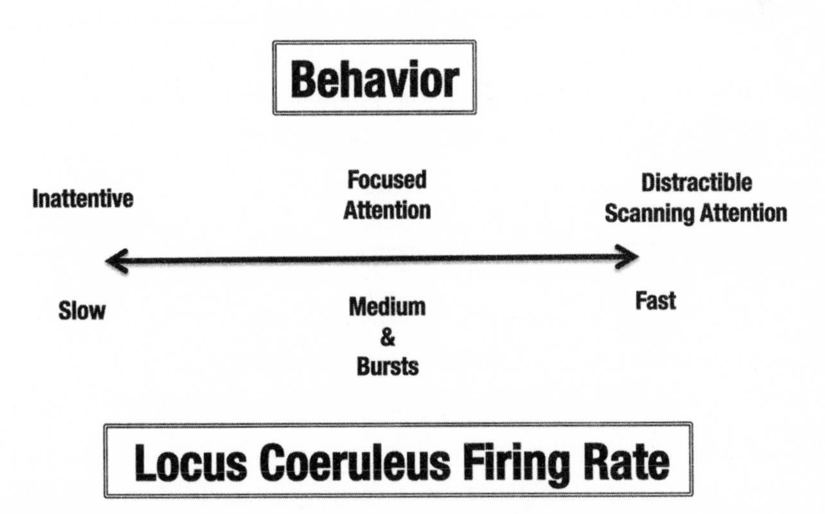

Figure 12: The behaviors seen at different locus coeruleus firing rates

As remarkable as it is that such a tiny cluster of nerves is tied so importantly to a person's behavior, the locus coeruleus does something even more startling. At slow and very rapid rates, neurons within the locus coeruleus fire independently, like stars thousands of light-years apart, twinkling in the clear night sky. But when the locus coeruleus fires at a moderate rate, the cluster radically changes its firing pattern. No longer do the neurons fire independently. When our attention is calmly focused, thousands of locus coeruleus neurons fire in synchronized bursts, pulsing nor-Adrenaline throughout the brain.

Locus Coeruleus *Erectus*

Now that we know a little bit about the locus coeruleus, let's look at what happened in this part of the brain of one of the *Homo erectus* on the savannah.

Lying back, drowsy from the meal, his locus coeruleus fired at a slow and steady pace. That quiet discharge rhythm was broken by the cry that announced the cat. The man was now fully awake, aroused to action. His locus coeruleus fired at an extremely high rate, bathing his brain in nor-Adrenaline. This corresponded to his eyes darting back and forth from the cat to the possible weapons and an escape route, assessing and reassessing the unfolding threat.

All the while his body's adrenaline surge was taking hold.

When the nor-Adrenaline system is active, the brain is awake; when another brain chemical, acetylcholine, predominates, the brain is asleep—in REM sleep specifically. The arousal's brainstem REM switch makes sure it is in the awake position in crisis.

The instant the troop's unspoken plans solidified, the man stood at the base of the ledge, perhaps with a stick he had grabbed. His unwavering eyes tracked each movement the cat made, anticipating

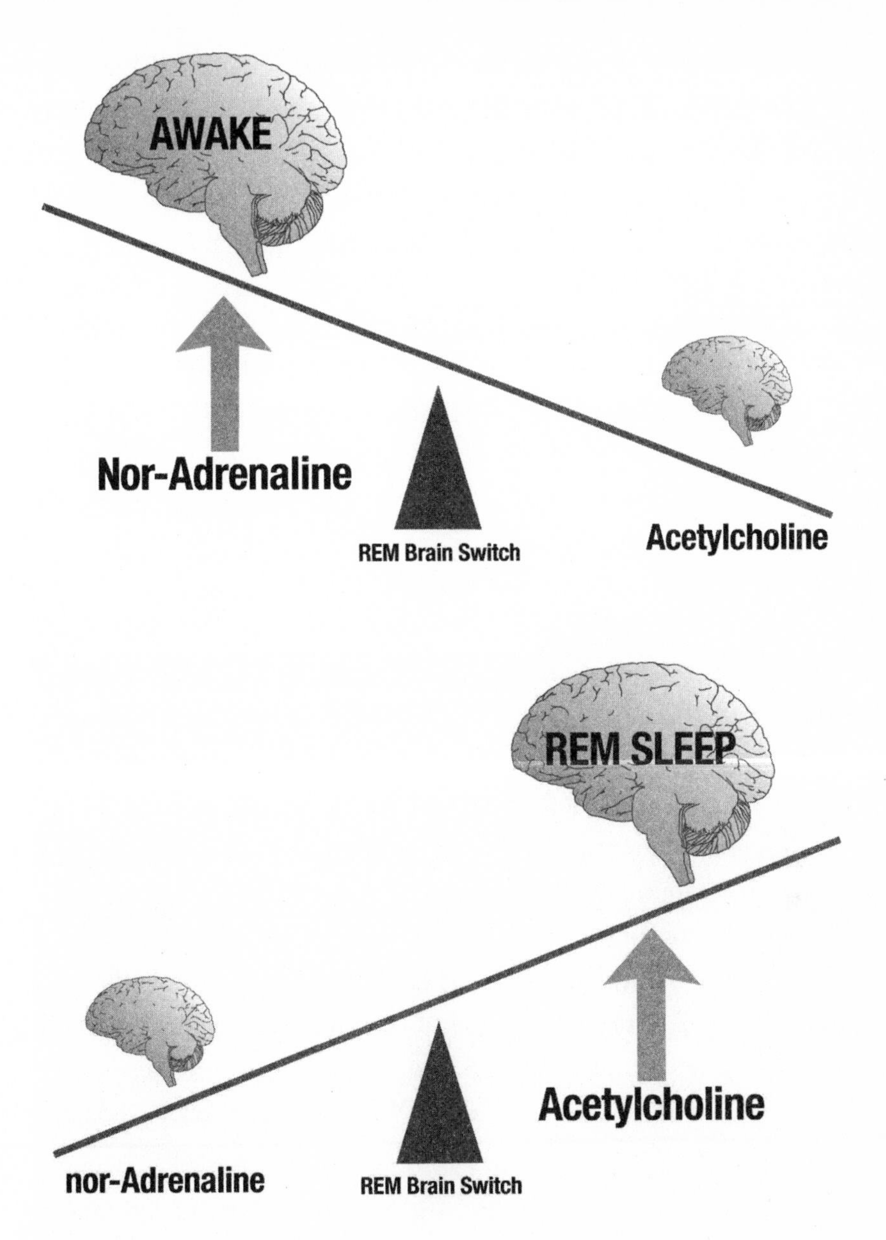

Figures 13a and 13b: The brainstem's reciprocal arousal system. When the nor-Adrenaline system is active the brain is awake (a), and when the acetylcholine system is active the brain is in REM sleep (b).

what it might do next. Before that instant, his locus coeruleus slowed to a moderate, synchronized, metronomic beat, and its bursts supplied his brain with nor-Adrenaline. The limbic system's hippocampus for distant memories and its prefrontal region for immediate working memory were blasted with nor-Adrenaline. It could be fatal if his attention broke for even a fraction of a second.

The locus coeruleus helped *erectus* orient his attention to the cat and climb the rock at exactly the right moment. It interacted with the brain network that focused attention on things outside his body. A critical component of the network coordinating our body schema is the right temporoparietal junction, which we have seen is important in out-of-body experiences.

When our hunter turned to finally make his way to the ledge's safety, the discharging pace of his locus coeruleus remained steady while his eyes focused on the moves necessary to make the climb. If his mind drifted to "I wonder if water will be nearby the next camp," he could be a *Homotherium*'s meal. To escape peril, *erectus* must orient and reorient his attention fraction by fraction of each second.

We've seen how nor-Adrenaline and the locus coeruleus are important to consciousness in response to danger, but before we look at the spiritual implications of this part of our neurobiology, let's look at the way our bodies and brains together respond to fear—the fundamental link, I think, to many of our spiritual experiences. From *erectus* on the savannah we jump forward in time to New Mexico and the dry country on the outskirts of Albuquerque, after I graduated from medical school.

An Unexpectedly Perilous Journey

I arrived in Albuquerque to begin my internship in July. One afternoon, when I was still getting settled, I bicycled out along the Rio

Grande to explore the city's outskirts. I pedaled out of town into the countryside. The bike trail meandered along the river. The summer sun was strong, but the high altitude made the ride comfortable, even for a Michigan boy. I took in the unfamiliar scenery, pedaling at an uncharacteristically leisurely pace. Suddenly, I heard a pop. It came from the direction of two men standing next to a white van about a hundred yards ahead of me. I heard another pop. This time there was no mistaking the sound—a gunshot. I took notice but wasn't concerned. When I was growing up, I had spent a lot of time out in the woods with firearms, and I assumed the two men near the van were target practicing. I kept pedaling and heard another pop. This time it was followed by a whizzing sound past my right ear. I stopped. Could they really be shooting at me? I looked around to see if there were other possible targets. I was in an empty field; not even a small brush stood within fifty yards. *Pop!* Dust kicked up near my front wheel. Incredible as it seems, they were shooting at *me*.

Strangely, I wasn't frightened. I wondered if they were trying to hit me or just scare me for their amusement. I didn't want to find out, so I turned my bike around and began to pedal away.

I had only gone a few feet when I looked over my shoulder—the van was moving, chasing me! I assessed the terrain. Between the bicycle trail and the dirt track for vehicles ran a steep arroyo. There was no way they could drive onto my trail, but that didn't matter. The trail, arroyo, and two-track paralleled one another for as far as my eyes could see.

I pedaled furiously, but they were quickly gaining on me. Then I caught a lucky break. Rounding a slight turn, I unexpectedly found myself out of their view for a few seconds. Tall grass covered a gentle embankment that sloped against the trail. A hiding place! I jumped off the bike and flattened myself against the ground, my face in the dirt.

I could hear the van approaching. Across the field was a trailer park. I knew I had only seconds to act if I wanted to bolt for the trailers, jackrabbiting across the field to safety before they could get the gun set. I felt confident: they had proved themselves mediocre marksmen. I was accustomed to long-distance running and felt pretty sure I'd get away. But I didn't want to leave my bike! I had very little money and the bike took me to the hospital each day.

The van cruised slowly past me and kept going. I stuck my head up when they were only twenty yards or so away. They could have seen me in the rearview mirror, but I had felt an irrepressible urge to see them drive away.

As it turned out, they kept going. When they were out of sight, I stood with my bike on the trail. The relief I felt was less than you might expect. But as I reflected on what had just happened, I began to tremble. Then I was shaking uncontrollably. This response stunned me. I could barely stand with all the adrenaline that was surging through me, but I forced myself to get on my bike and pedal out of there. I wasn't biking hard, but my heart was racing, pounding in my chest; my breathing was shallow and quick. My body was prepared to move and move fast, but to where? My mind was clear: I didn't feel panicked or even particularly afraid, yet my body was primed for action.

In the episode on the savannah, we saw how the brain primes itself for action. But let's look more closely at how brain and body interact in moments of crisis.

Two points really stand out when I think of the shooting incident today. The first was my lack of fear. While I was being pursued, I felt calm and dispassionate, almost disconnected. I knew what was happening to me, and I was not so far removed that it seemed unreal or as if I was watching a movie about somebody else. I felt present but no more mentally agitated than if I had been sitting at home reading a book. My subsequent trembling surprised me.

Second, during my escape, time slowed to a crawl. Just like *erectus* on the savannah, I scanned my environment for an escape route. Thoughts and ideas came in rapid succession, and I felt that I had all the time in the world to make my getaway.

In retrospect, I wondered if the slowing of time enhanced the acuity of my senses and my ability to assess danger and make decisions; I realized my detachment and speeding thoughts were my brain's survival reactions.

It wasn't until recently that I understood the brain wiring that produced this state. I was reminded of the surge in my adrenaline that followed the chase when Cliff (the rehab doc involved in the lawsuit) related his spiritual experience to me, although there were significant differences between his experience and mine. Cliff was emotionally but not physically endangered, and his adrenaline surge was immediately followed by a spiritual epiphany. I realized I was endangered yet never considered my experience as spiritual. Still, as superficially different as our experiences may seem, I think they have important similarities. For both Cliff and me, the agitation came in the aftermath, presumably because the adrenaline surge hung on while we no longer had a crisis to focus our attention.

It is these similarities that formed the basis of my theory of what happens in the brain during a near-death experience. So far all the discussions about near-death experiences had ignored the key instincts that guide us through crisis and keep us alive. The neuroscience of near-death experience (such as it is) had focused exclusively on other parts of the brain. I wanted to understand how the brain and body's survival instincts come into play during near-death experiences.

With this in mind, I traveled to Harvard's Countway Library of Medicine to look for clues on what might be happening near death, from the first man to understand the body's survival responses.

"Fight or Run"

In the protected archives of the library's special collections lies the daily journal of one of science's greatest physiologists, probably William

James's most accomplished student, and from a scientific perspective arguably a thinker more influential than James himself. Walter B. Cannon was stocky and average in height; his Midwestern clothes were neat but contrasted sharply with Harvard's well-groomed elite. He was devoted to his family, magnanimous to his professional rivals, and an early, active force in protecting Jewish scientists in Europe during the Nazi era.

James and Cannon respected each other, but they had a complex relationship. Cannon spent much of his career arguing against the "James-Lange" theory of how the body expresses emotions, especially fear. The difference between them was kind of like the argument about which came first, the chicken or the egg. James held that emotions depend on our body's physical reactions. You see a bear, your heart thumps, you experience fear: the fear comes from the sensation of your heart thumping. Cannon took the opposite position: you see the bear, the brain feels fear, your heart thumps.

Cannon's theories have turned out to be closer to the truth. Patients whose autonomic nervous systems have been destroyed, with hearts disconnected from their brains, can still feel fear. The bear can not set their heart to thumping, but they can still *feel* afraid of bears.

For many years, portraits of Claude Bernard and Charles Darwin hung in Cannon's office. In 1865, Bernard introduced the idea of the *milieu intérieur,* referring to the stable environment the body provides for each of our cells. Cannon dynamically expanded that idea to include chemistries like blood sugar and oxygen, which, he later proved, were also important during threat and crisis. Coining the word "homeostasis," Cannon showed how the body normally kept nutrients like sugar and oxygen balanced within a narrow range. He examined those physiological mechanisms at work when a crisis threatens homeostasis. Such threats include bleeding, pain, and fear.

Cannon discovered that the adrenal glands secrete a hormone, which he called adrenaline. Adrenaline acts in concert with the sympathetic nervous system based on nor-Adrenaline, the nerves' and brain's form of adrenaline. Nor-Adrenaline is slightly and insignificantly chemically different from the body's more widely known adrenaline.

The adrenal gland and its companion, the sympathetic nervous system, are the body's first defense against attack. Confronted with danger, the body must be instantly primed to struggle or flee. The lung's air passages open to receive air and enrich the blood with oxygen. The heart races and blood pressure rises to deliver the blood where it is most needed—the brain and muscles. Simultaneously, adrenaline liberates sugar from the liver into the blood. Dilated arteries supply muscles with the nutrients needed for action. Pupils dilate, letting in light, aiding our vision. Sweat dissipates the heat created as our muscles burn up energy. The body stops producing saliva so precious water can be used to expand our blood supply and make sweat. Blood is diverted away from organs that are not immediately needed, like the gut and skin.

We've all experienced this adrenaline surge. It's the basis for great feats of strength in times of crisis. It's unleashed on the battlefield as well as the gridiron.

Cannon expanded upon Bernard's idea of balancing the body's *milieu intérieur* also by connecting it to Darwin's ideas of evolution. In the struggle for survival, natural selection favors the animal whose adrenaline surge enables it to meet the demands of crisis.

Pairing this surge with the survival reaction of "fight or flight" came to Cannon rather suddenly. Cannon's small daily journal, begun in 1911, spans four years. One of the earliest entries (January 20, 1911) ends with a rare and extraordinary exclamation: "Tried experiment on rabbit—but no success. Got idea the adrenals in excitement serve to affect muscular power and mobilize sugar for muscular use—thus in wild state readies for fight or run!"

Cannon's idea of adrenaline and the sympathetic nervous system priming a person for danger has stood the scientific test of time.

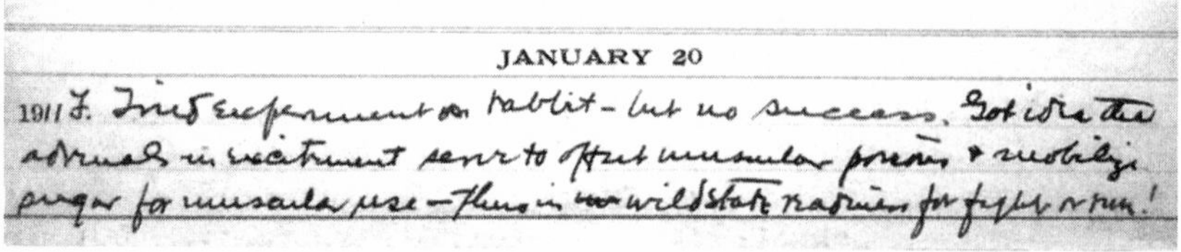

Figure 14: Walter B. Cannon's journal entry recording when the momentous idea for "fight or run" first came to him.

We now understand that the adrenaline surge that primes the body for fight or flight has been fundamental to our evolutionary development from fish to human beings—its circuitry resides in the evolutionarily oldest and most primitive region of our brain.

Such a powerful system must not go unchecked. The sympathetic nervous system that acts with the adrenal gland is one of two great nerve and chemical systems that Cannon said stood in balanced opposition to each other. The adrenal gland and sympathetic nerves together act against their opposite—the parasympathetic nerves. In the call to action, the sympathetic nervous system dominates; the parasympathetic takes over when we're resting. The parasympathetic nerves use the chemical acetylcholine to communicate their impulses to other tissues, for example to slow the heart rate or speed the movement of food through the gut.

The sympathetic and parasympathetic combine to form the autonomic nervous system. The autonomic nervous system controls the internal organs or "vegetative" functions, determining the internal milieu. It is "autonomic" because it governs our organs without needing, and rarely benefiting from, our conscious will. We do not have to consciously will each heartbeat or breath—that is the job of the autonomic nervous system. The fight-or-flight response is a fundamental mind-body link—a way that mental stress acts on the body.

As we have seen, the autonomic nervous system is the core of the

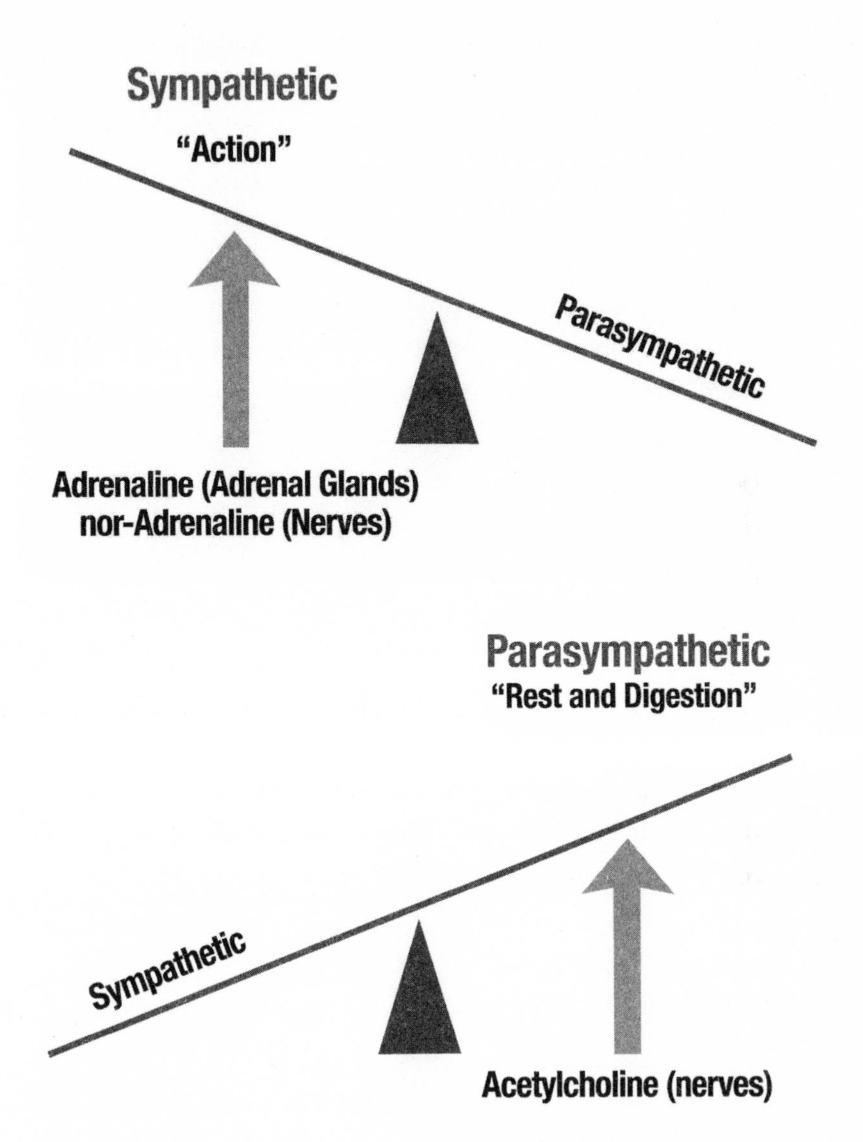

Figures 15a and 15b: The sympathetic and parasympathetic nervous systems act in opposition upon our organs balancing our body's internal milieu. Nor-Adrenaline is the same chemical the brain uses to maintain waking consciousness. Acetylcholine is the same chemical the brain uses for REM sleep.

body's fight-or-flight reaction. The sympathetic and parasympathetic poles act in yoked opposition to each other. When one is up, the other is reciprocally down. In crisis, the sympathetic nervous system with its adrenaline and nor-Adrenaline is up, while the parasympathetic system with its acetylcholine is down. This reciprocal system, controlling heart and lungs, extends directly upward to the brainstem arousal system and the switches controlling consciousness.

Brain and body systems mirror each other and are chemically interlinked. The brain is not only a producer of Shakespeare's plays but a nuts-and-bolts organ of nerves that always has to work in lockstep with the body, even before our forerunners left the sea to crawl on land. As we saw earlier, brain chemistry developed as a direct extension of body chemistry.

Voodoo Death

When I was at Harvard reading on Cannon, one of the library's staff laughed when he brought me Cannon's journal. While locating the journal, he had stumbled across a box of materials labeled "Voodoo Death." It was right up my alley, of course. I already knew about the box's contents. "Voodoo Death" was the title of an article that Cannon published in 1942 in the journal *American Anthropologist.*

It may seem strange that such a highly regarded scientist would trouble himself with a fringe topic in a journal far outside his field, but there was a method in his madness. Cannon understood that anthropologists knew of cases where a spell or sorcery was thought to have brought about death. He wondered if "an ominous and persistent state of fear can end the life of a man." With a physiologist's eye, he sifted through reports (many of them contained in that box), hunting for clues as to how the fight-or-flight reaction could be fatal.

For Cannon, voodoo death was not a joke. Augury or hexing from afar *could*, truly, end someone's life. He based this on what he knew about fight-or-flight. The voodoo victim was like a trapped animal, convinced he or she was hopelessly doomed. The victim would stop eating and drinking, and Cannon hypothesized that the unrelenting adrenaline surge "to obvious or repressed terror" eventually drew precious fluids out of the blood, leading to the unsustainably low blood pressure of shock, which is exactly what had happened to soldiers with severe blood loss that he had treated when he served as an army physician during World War I.

Cannon discovered the physiology of shock that caused many battlefield deaths, and it drove him to devise methods to resuscitate the soldiers by bolstering their circulation with fluids, minerals, and baking soda! Cannon considered it "a red letter day in my surgical experience" when he snatched a wounded soldier from death for the first time. His ideas saved countless lives in World War I and afterward became the basis for the treatment of trauma victims that persists even today.

This is the same kind of shock, by the way, that Jan entered when her husband's gun accidentally discharged, which culminated in her near-death experience during surgery.

Cannon published his ideas about voodoo death as an appeal for anthropologists around the world to make observations in the field about adrenaline surge. Yet on the final point of how the autonomic nervous system causes death, Cannon was only partially right. It is clear that people can be frightened to death through the adrenaline surge, but the collapse of blood pressure does not happen quite in the way he envisioned.

We still don't know the full physiological story, but we do know that voodoo death ends at the heart. The brain, through the adrenal glands and sympathetic nervous system, can make the heart beat irregularly and arrest. A tremendous adrenaline surge can destroy heart cells directly. Sudden and severe emotional stress can stun the heart, causing its cells to die in stiff contraction in a kind of severe cramp, even to the point of turning the heart hard as a stone. The brain and adrenaline surge account for sudden and otherwise unexpected death in conditions such as stroke, seizures, and head trauma, and during terrifying natural disasters such as earthquakes and hurricanes. This seems to be what happens when people accidentally die from being stunned by electronic weapons such as a Taser.

In short, you can die of fright. The adrenals, fight-or-flight, and the sympathetic nervous system kick into overdrive and run in their highest gear for too long. We literally burn out—driven to physical exhaustion and collapse.

Darwin's Fall

In his autobiography, Darwin wrote: "I have heard my father and elder sister say that I had, as a very young boy, a strong taste for long solitary walks; but what I thought about I know not. I often became quite absorbed, and once, whilst returning to school on the summit of the old fortifications round Shrewsbury, which had been converted to a public foot-path with no parapet on one side, I walked off and fell to the ground, but the height was only seven or eight feet. Nevertheless, the number of thoughts which passed through my mind during this very short but sudden and wholly unexpected fall, was astonishing, and seem hardly compatible with what physiologists have, I believe, proved about each thought requiring quite an appreciable amount of time."

I don't know what Darwin thought the physiologists of his time had proved. What I do know is that psychologists today have taken an intriguing look at how we experience time during crisis, and this directly relates to the way people see their lives flash in front of their eyes near death and how we may experience time during other types of spiritual experience, which is one of the reasons I think these experiences are often referred to as "realer than real."

What happens in the brain during times of crisis that makes us feel that time slows down? Did Darwin's thoughts actually speed up when he fell? Does the brain act like a high-speed camera when we're in danger, sharpening our perceptions, enabling us, in the extreme, to watch the beating of a hummingbird's wings in flight? Enhanced vision would give us a powerful evolutionary advantage in the struggle for life. We could better react to the rock being hurled at our heads or predators leaping at our throats. When I was being chased, time slowed, or my thoughts sped up, and I had no trouble making snap decisions that kept me from getting shot.

Psychologist Chess Stetson and his team at Baylor University in Texas wondered if brain processing speeds up when we're frightened. They took several subjects to the top of a special tower to free-fall backward one hundred feet into a safety net below. They strapped a special wristwatch on each subject's wrist, which flickered a number at intervals slightly faster than the subject could read when calm and on the ground. The investigators reasoned that during the two-and-a-half seconds of free fall, fright might accelerate the brain and their subjects could read the number on the watch. But failing to read the number could result from something that has nothing to do with speeding up brain activity. The subjects have to keep their eyes open during the entire frightful fall if we expect to draw valid conclusions. To make sure eyes remain open, a researcher had to fall alongside each subject.

As they plummeted, none of the thirteen subjects could read the flickering number. It turns out that fear does not speed the brain up and let us see events more clearly. This is disappointing in a way. The neurologist in me would like to see the brain capable of faster processing in crisis. Instead, what this experiment suggests is that we *perceive* time slowing or our thoughts speeding up when we're in danger.

The change in the subjects' time perception stayed with them even after the fall. Before jumping off the tower, the subjects started and stopped a timer as they *imagined* someone else falling to the net below. After they were done with their fall, they replayed it in their minds, gauging its duration with a stopwatch. They consistently judged their falls to have lasted more than a third longer than the imagined fall of someone else.

The way we perceive time is complex and incompletely understood. During a crisis, the brain has to rapidly synchronize diverse sensations: sight, sound, body position, and motion. The brain does not process each sensation in the same way: after all, knowing the position of your right foot is very different from the pathways the brain uses to process electromagnetic light waves. Different sensations are processed at different speeds before being transferred to memory. Calibrating these sensations into a kind of unity is a critical brain function—when things happen in time forms the basis of how we experience our larger sense of causality. Synchronizing sensations may be tied to how the brain synchronizes our experience, and except in rare instances, it is the reason we experience ourselves as inside rather than out of our bodies.

Stetson and his colleagues surmised that time perception and memory are closely intertwined. The perception of time slowing is not the true slow-motion experience we see in the movies. It has to do with the way we form *memories* in the limbic system.

The way the brain creates and retrieves memory is a fascinating topic, and I don't want to delve too deeply into it here. But how the brain forms memory when we experience fear is crucial to the near-death experience and needs some explanation.

Fear and the Limbic System

When we are facing a frightening experience, the hippocampus in the brain's limbic system is busy forming rich memories. The more memories stuffed into a second, the longer that second seems to be, reasoned Stetson's group.

The hippocampus is a key limbic structure for the memory and emotions that become the foundation of spiritual experience. The hippocampus is closely tied to the amygdala, a complex concentration of nerve cells (nuclei). The amygdala relays information: it's a kind of switchboard, linking the lower brainstem, with its connection to the heart and lungs, to the thinking and decision-making process of the cerebral cortex, including the hippocampus. The amygdala processes—coalesces, packages in meaningful ways (we're actually not sure exactly what it does or how it does it)—the sensations from the brainstem that underlie fear and anxiety, when our hearts are racing or our lungs hunger for air, the vital information from the internal milieu. These are the sensations we depend on for survival and include sound, sight, feeling, and taste, which reach the amygdala from other brain regions.

Remember Dr. Penfield's patient "V," whose seizures were prompted by smells and accompanied by fear? His seizures arose from the limbic system and undoubtedly involved the smell and fear circuits connected with the amygdala and hippocampus.

Because the hippocampus is important to the way the brain creates memory, it is also important to the way we experience time. We haven't found a single unified brain clock that measures time second to second, but we do know that from very early on in our evolution our brains have been wired to rapidly detect and react to danger. This is true in very early forms of life. When you are about to eat or be eaten, fractions of a second matter.

Evolutionary forces have refined the body and brain for survival. Primitive survival reflexes and instincts have stayed with us and have been elaborated in very human ways. We react first and exercise what free will we have later. A simple example of this is the reflex in the spinal cord that allows you to quickly withdraw your bare foot when you step on a piece of broken glass.

A left and right limbic system is mirrored on each side of the brain. If both are lost, so, too, is the ability to recognize fear on other people's faces. At the same time, we can still register fear subconsciously. Seeing fear on the face of someone else is very important when we face danger together as a group (as did the *erectus* troop confronting the cat). We can have a fearful body disconnected from feeling fearful, as I learned when I was shot at in New Mexico.

It should come as no surprise that the brain makes and retrieves memories during the near-death crisis even if one does not feel fear. My bike ride indicates that we don't need to feel fear in order to experience a fight-or-flight response. Time slowed during the incident and my memory was vigorously active.

Memories made and recalled during experiences when we perceive danger can be critically important. If we survive, it is good to remember what helped us get out of the crisis, or remember how we got there in the first place! (On the other hand, a flood of crisis memory can disable someone with posttraumatic stress disorder.)

The amygdala has extensive connections with the nearby hippocampus in the temporal lobe, where we form the synapses of memory. The hippocampus is so named because it is shaped like a seahorse (from the Greek, *hippos*, for horse). Both the amygdala and the hippocampus are integral to the autobiographical memory so important to surviving and avoiding situations that evoke fear. The memories we form when we're afraid are strong and often lifelong, even if we lose the original fear. Many adults vividly remember being afraid of the dark as a child, even if the dark no longer holds the same fear for them now.

The tight interconnection between the amygdala and the hippocampus (and other brain structures) is why emotional experiences are memorable. Not surprising, the amygdala's role takes place with an interplay of adrenaline, acetylcholine, and stress hormones. It's interesting that fear emotions (anxiety, terror, worry, horror) and the enhanced memory of frightening events may take place in the amygdala *below* the conscious level, which makes sense because the amygdala must detect danger rapidly. Thinking involves coordinated interactions among huge numbers of neurons and synapses, which, in turn, take precious time. This is why we have reflexes in the lower parts of our brain, below the thinking cortex.

I half-jokingly tell my students that one of the amygdala's roles is to use fear to bolster memory (or vice versa). So they should be grateful when I bring fear to their lessons: it's for their own good. They never seem convinced of that reasoning, but they always remember it.

By now we have explained much about how the brain responds to crisis and processes the feeling of fear involved in near-death and some other types of spiritual experiences. Yet we are still far from explaining the bliss that accompanies so many near-death experiences. How does the brain move from terror to spiritual bliss? Few have described this transformation as well as Fyodor Dostoyevsky.

A Revelation in Semenovsky Square

At four in the morning on April 23, 1849, in St. Petersburg, a young Dostoevsky was awakened by an officer of the tsar's secret police and dragged off to prison, charged with being part of a group of progressive intellectuals who were writing and speaking out against the bondage of the serfs in tsarist Russia.

Dostoevsky was interrogated and thrown in a cell, where he languished for two months. Then, early on the morning of December 22, guards led Dostoevsky, and other prisoners similarly charged, in their light clothes, to a line of empty carriages, which were waiting outside in the biting cold. After a thirty-minute ride, the carriages stopped and the prisoners emerged in Semenovsky Square. They were surrounded by armed troops. A crowd had been gathered. In the square's center, Dostoevsky saw a newly constructed scaffold, ringed with black crepe. A priest led the prisoners silently up to the scaffold platform, where they were all sentenced to death and given a funeral shroud to wear. The priest, a Bible in one hand and a cross in the other, urged them to repent. He moved from prisoner to prisoner, and each kissed the cross, which the priest pressed gently to the prisoner's lips. Even the confirmed atheists took part in the ritual.

The first three men in line were taken off the platform and tied to nearby stakes. Two had their heads covered, while the third looked defiantly at the soldiers who had formed a firing squad. Dostoevsky was standing waiting in the next group of three, preparing for his demise, when the drums rolled. Dostoevsky had been an officer in the army and instantly knew that the drums were beating retreat and that his life would be spared. Which is exactly what happened. An aide-de-camp galloped into the square, bearing the tsar's true sentence: Dostoevsky was sent to a prison in Siberia.

During this mock execution, one of the men who had been tied to the stake in front of the firing squad, his mental state already fragile from imprisonment, had what Dostoevsky described as a nervous collapse. Dostoevsky reported that the man's face was ashen, drained of blood, which makes me think he may have been close to fainting. Apparently, this man never emotionally recovered from the experience.

Dostoevsky, on the other hand, facing impending death, had a very different response. He was spiritually awakened. As soon as he returned to his cell, he hastily penned a letter to his older brother before the authorities whisked him off to Siberia. In it he wrote that facing death had given him a new grasp of life. He was in the throes of ecstatic revelation, awakened to the dazzling truth that life itself is the greatest of all blessings and each of us has within us the power to turn each moment into an "eternity of happiness." Standing on the scaffold, awaiting the firing squad, Dostoevsky felt the urge to forgive and be forgiven. Embracing others with unconditional love and forgiveness as he faced death became for Dostoevsky the supreme human virtue, a conviction he would always carry with him. Years afterward he would tell his wife: "I cannot recall when I was ever as happy as on that day."

Dostoevsky was steeled by his spiritual renewal. "I am reborn in a new form," he wrote in the face of his impending Siberian imprisonment. From this point onward, religious faith became central to him and provided him with literary material and vital artistic energy for the rest of his writing life. In his last novel, *The Idiot*, written twenty years after his mock execution, a mature Dostoevsky described his transforming ordeal in Semenovsky Square through the eyes of his Christ-like hero, Prince Myshkin, who recounted the thoughts of a similarly condemned man:

> "He had only five minutes more to live. He told me that those five minutes seemed to him an infinite time, a vast wealth; he felt that he had so many lives left in those five minutes that there was no need yet to think of the last moment, so much so that he divided his time up. He set aside time to take leave of his comrades, two minutes for that; then he kept another two minutes to think for the last time; and then a minute to look about him for the last time. He remembered very well having divided his time like that. He was dying at twenty-seven, strong and healthy. As he took leave of his comrades, he remembered asking one of them a somewhat irrelevant question and being particularly interested in the answer. Then when he had said good-bye, the two minutes came that he had set apart for thinking to himself. He knew beforehand what he would think about. He wanted to realize as quickly and clearly as possible how it could be that now he existed and was living and in three minutes he would be

> *something*–someone or something. But what? Where? He meant to decide all that in those two minutes! Not far off there was a church, and the gilt roof was glittering in the bright sunshine. He remembered that he stared very persistently at that roof and the light flashing from it; he could not tear himself away from the light. It seemed to him that those rays were his new nature and that in three minutes he would somehow melt into them."

To a neurologist it is clear that as Dostoevsky stood on the scaffold, his adrenaline surged and his limbic system seared autobiographical memories into his hippocampus. Making so many memories explains why "those five minutes seemed to him an infinite time."

Unlike his ashen companion, there is no indication that Dostoevsky was about to faint and deprive his brain of blood as we commonly see in near-death experiences. So how did Dostoevsky's brain transform from fight-or-flight and fear to the rapture of a dazzling truth? In his "last" thoughts we see a common near-death theme of expansiveness and love: "Only then did I know how much I loved you my dear brother," he wrote in his letter. In those minutes he concentrated on everything he held precious. "Life is a gift," he wrote his brother—and, in the end, he was rewarded with the miracle of life. I'd like to suggest, although it may raise some hackles, that how the brain handles rewards, big and small, is where we should look for the lasting bliss that accompanies various types of spiritual experience.

The Brain Receives Its Divine Rewards

As we've seen, during fight-or-flight events the brainstem sends a torrent of information along to the amygdala, which relays that information to other regions of the limbic system, including the prefrontal region. This part of the limbic brain underlies our ability to feel good about something. Within the limbic system are two closely related regions: the orbital prefrontal and the medial prefrontal regions.

The medial prefrontal region is at the top of the limbic system, governing the heart, lungs, gut, and sweat that we experience during emotions. A strong relationship with the amygdala enables the medial prefrontal brain to make sense of our visceral responses, setting the moods that guide our actions and choices.

If the medial prefrontal region is damaged, it leaves people without the normal automatic visceral responses—the feeling of fear when they really should be afraid, for example. They become severely disturbed and can become self-destructive or deranged, even though their rational intelligence is unimpaired. Emotional intelligence influences all types of decisions, from our family life to our business dealings. This is partly why the rod passing through Phineas Gage's skull, damaging his medial prefrontal brain, led to his downfall.

People like the injured Gage will choose immediate reward and show no regret for their decisions. The research team of neurologist Antonio Damasio, then at the University of Iowa and specializing in behavior, has extensively studied Gage's brain injury. He believes these unfortunates like Gage have lost the "somatic marker" that gives our subconscious visceral warnings to avoid certain behaviors.

A second limbic network with strong amygdala connections is the orbital prefrontal. Here the brain assigns emotional value to pleasure and rewards. Once the orbital prefrontal brain coins the value of something (be it high or low), it transmits that value to other brain regions so we can make the right (or wrong) decisions. *Erectus* may have discovered that a certain tuber was nourishing, and their brains may have given it a high value. The orbital prefrontal primes other brain regions, like those serving vision, so *erectus* could easily see the leaves of the bush and find it again.

Much as we may rebel at treating something as sublime as spiritual experience as the crass product of the brain's reward system, the fact remains that at a very basic level experiences that spring from brain processes are reward-based. Our brains evolved with rewards having high survival value: the rewards of food and sex, for example, which are often very much on our minds. The orbital prefrontal brain receives sensory input from smell, taste, vision, and bodily sensations and blends these sensations to give us the pleasures of food.

This is how our evolutionary forebears enjoyed the pleasures of food: because the centers of taste and smell strongly connect with the orbital prefrontal region. For *erectus*, the musky smell and taste of the freshly killed antelope enriched his orbital prefrontal brain on many good days, including that one of anecdote.

The orbital prefrontal region responds to food temperature, texture, and flavor, like creamy or sweet. It must get heavy use as babies nurse. When we smell Mom's apple pie baking and switch to our best behavior to get a piece, the brain's reward system has come into play. For some people the reward is chocolate. Just the sight of chocolate is enough to make the orbital prefrontal brain of chocolate cravers "light up" on an MRI.

Although the pleasures of food are the foundations of the orbital prefrontal area in other animals, the orbital prefrontal regions and its pleasures (or rewards) extend to virtually *every* area of modern human life. The reward system guides our behavior, and it has a dark side. It's the basis for our addictions, not just to chocolate but to heroin, cocaine, and gambling.

If we accept that the brain participates in spiritual experience, then the orbital prefrontal must be responsible for giving us a glimpse of the rewards that will be ours when we go to heaven or reach enlightenment. It transforms near-death experience, in fact *all* spiritual experience, into a reward of the highest priority. Dostoevsky walked through a spiritual doorway that included his orbital prefrontal brain, experienced the transcendent, and it transformed his life.

Attention to Reward

The orbital prefrontal is the highest tier for making decisions based on rewards. Once something (like food) has been assigned a reward value by the orbital prefrontal brain, it's disseminated to other areas of the brain as well.

The orbital prefrontal region mints the common currency that represents rewarded value throughout the brain. In turn, other brain regions occupied by thinking and paying attention influence the orbital prefrontal and the reward system. The locus coeruleus and the orbital prefrontal region are robustly connected. After all, we have to be paying attention if we expect to get rewarded. A baseball hitter focuses on the baseball hurtling toward him. The sound and feel of getting good wood on the ball reinforces the pleasure of seeing the ball's arc as it flies into the stands for a home run.

The orbital prefrontal brain and vision are also connected, so we can see the objects participating in our desires. As social creatures, we recognize the smiles and frowns of others that direct our behavior to the social rewards we seek.

Words and thinking alone can alter the reward value assigned by the orbital prefrontal cortex. It is the special intelligence of our species that rewards can be abstracted and rapidly updated, and our rewards can be several steps removed from what drives our behavior. Everything from a virtuous deed to a good cup of coffee will be represented in the prefrontal region after it leaves our immediate senses. The massive size and complexity of the human prefrontal region for reward, like the left hemisphere for language, is what most distinguishes us from the apes.

The orbital and medial prefrontal cortex links goals and emotions, helping us prioritize competing rewards. The thinking goes that rewards with some kind of emotional pleasure or satisfaction reinforce survival behaviors. Our genes may also, of course, help us select the rewards that favor survival. When these goals have been rewarded consistently over an extended period of time, the reward may not be immediate but the behavior becomes a habit—the batter routinely making his way to the batting cage for practice, for example.

The reward system comes into play in our everyday life. When we go shopping, we often strive to make buying decisions consistent with how we see ourselves. The orbital prefrontal brain assigns the reward value to our buying options. It is the medial prefrontal area that weighs these values to make decisions in a timely fashion.

Although the orbital prefrontal brain assigns values for the brain and influences our life's choices, if we have free choice, it may be exercised within the medial prefrontal brain.

We are not at the complete mercy of our instinctual reactions in our fight or flight. The medial prefrontal region has evolved to allow us to rein in the visceral fight-or-flight reflexes while still allowing for quick "gut" reactions. Our medial prefrontal brain can call upon a broad range of emotions that promote right and suppress wrong behaviors. We use our medial prefrontal brain to weigh our visceral reactions as we make our choice. Once they are weighed, the medial prefrontal area brings other brain regions in line with the choice we've made.

It could be the medial prefrontal region that helped *erectus* suppress his urge to immediately flee when he first sighted the cat.

What if free will, like spiritual free will, is not a rational choice, like selecting the best television for the money? What if spiritual free will, in whatever form it exists, is a limbic emotion, a feeling, or nothing more than a hunch?

Could the medial prefrontal brain, critical to our visceral responses, also be where some of us choose to be born again Christians, Buddhist monks, agnostics, or atheists?

For many people, spiritual experience is one of the most, if not *the* most rewarding of life's experiences. As a neurologist, I have no doubt that Dostoevsky's limbic prefrontal reward system must have been permanently changed that December morning during his mock execution, and that change influenced the course and the caliber of his writing.

So Where Have We Been?

Up to this point we have seen many things contribute to the experience of being near death. Low blood flow reaching the brain from fainting or cardiac arrest leads to many features of near-death experience. If the temporoparietal portion of the brain shuts down, we have an out-of-body experience or have "sensed presence." When blood flow is cut off to the eye as well as the brain, we experience tunnel vision.

Fright alone can cause a near-death experience. The thoughts and feelings that came to Dostoevsky when he faced the firing squad are themes I commonly see as part of the near-death narrative. We have seen how these themes are intertwined with the limbic system and its reward components.

Now we turn to an essential element of these experiences, the mechanism that can bring the qualities of light and narrative. It must be a system that can be tied to fainting and cardiac arrest, the consciousness switch and the limbic system. What happens when our eyes move rapidly while we sleep? Is REM-state consciousness key to unlocking a spiritual doorway in the brain or just a nice idea that heads nowhere?

7

THE BORDERLAND OF DREAMS AND DEATH

WHAT MAY COME?

> "To die,—to sleep;—To sleep! perchance to dream! Ay, there's the rub; For in that sleep of death what dreams may come."
>
> —Shakespeare, *Hamlet*, Act III, Scene I

To understand the last piece of how the brain functions during near-death experience let's briefly return to *erectus* on the savannah and examine a very different scenario from the one we imagined at the beginning of Chapter 6.

> The saber-toothed cat crouches at the edge of the camp, fangs bared, poised to spring. The last two hunters stand below the other members of the troop, who are now safely perched on the overhanging ledge. One of the men who remains on the ground turns and scrambles upward, dislodging pebbles and small stones, which tumble around the last member of the troop who remains below. The man on the ground hears the rocks fall

> and reflexively looks for their source above. The big cat, which has been watching him, well, like a cat, notes his momentary distraction. The man fleeing up the rock to safety has provoked the cat to attack and she strikes at the closest *erectus*, the last one below the ledge. His eyes are diverted, and in three bounds the cat is upon him and sinks her teeth into his throat. She shakes her head and her curved incisors slash his jugular vein, spilling *erectus*'s lifeblood onto the dirt. Within seconds his blood pressure collapses and his consciousness fades into the borderlands as he slips into shock.

What happens to *erectus* during those last seconds of life? We have seen how his arousal system would have activated his hippocampus and memory retrieval and how time for him would have seemed to slow. He was hyperalert, hyperawake, aroused to fight or flight. His brain and body were still in that state when the big cat pounced and delivered the coup de grâce. As his blood pressure dropped and the blood drained from his eye, he might have had tunnel vision and, perhaps, an out-of-body experience. But last, and perhaps most significantly for a near-death experience, as he was dying, a mechanism might have been triggered that is so powerful it completely silences the locus coeruleus—the only time in our routine lives when the ancient metronome stops and our locus coeruleus is silent. With this silence might have come a heavenly light, taking him to the borderlands of consciousness and a place of vision and splendor in the waning moments of life.

We are now ready to talk about the last brain mechanism that contributes to the near-death experience—that mechanism that allows us to see things in the dark of sleep, that moves our eyes behind our eyelids as if we were watching the most compelling show on earth: REM consciousness. It is a show that is a mystery at its core—we don't even know why we enter REM consciousness and dream.

Death and Sleep

In his soliloquy that begins, "To be or not to be," Hamlet longs for death, absolute annihilation, the sleep of supreme nothingness. But then he has second thoughts—who knows what dreams the death of deep slumber will bring each night? Likewise, Hamlet is wary of what may or may not greet him when he crosses into "the undiscover'd country from whose bourn no traveller returns."

Shakespeare reaffirms a link between sleep and death that people have made throughout history. But only now has science begun to have the tools to describe biologically what this link might look like in the brain.

Why would a neuroscientist expect the dying brain to shift consciousness to sleep, just before all consciousness ceases? Turning on REM consciousness in crisis seems at first to have little biological or survival value. How could this consciousness possibly have come to the aid of *erectus* in the jaws of a predator? That question will take on a new perspective after we know a bit more about REM consciousness and how the brain brings it about.

To appreciate how the brain switches to REM consciousness as death approaches requires us to take a closer look again at the oldest parts of our brain, in particular the arousal system and the brainstem. The role of the brainstem is counterintuitive to the neurologist who is looking for the roots of near-death experience in the brain's more recently evolved cerebral cortex.

Lightning Bolts in the Brain

Brain activity never really stops in our natural lives (unless we're immersed in ice water). Even in sleep the brain is active and does not shut down as much it does when we're under anesthesia. Our brains are especially active during REM sleep. Neurologists refer to REM sleep as "paradoxical sleep" because from the standpoint of EEG-measured brain waves there is little difference between REM sleep and wakefulness.

When we dream, powerful electrical waves start in the pons (brainstem) and radiate upward to the geniculate (visual thalamus) and then to the occipital (visual) cortex. These waves, PGO waves for short, are figurative lightning bolts, electrifying the visual brain and creating the visions of dreams. Wave after wave ripples through the visual brain for as long as we are in REM. If you look closely at someone in REM sleep, you can see his or her eyes darting from side to side beneath closed and twitching eyelids—a sure sign that the person's visual system is active.

We do dream in drowsiness and non-REM sleep, but it is during REM sleep that we spend most of our dreaming time. Certainly, it is during REM sleep that our dreams are the longest and most fully developed.

In slightly more than a decade, brain scans (MRI and PET) have given us images that have revolutionized our understanding of brain activity during REM sleep. We can see which brain regions are active or inactive, which has explained a lot about our dreaming experience.

Dreams are primarily visual. Touch, smell, taste, and sound are less important than imagery. Dream emotions, especially fear, are connected to the activation of the limbic amygdala and the orbital frontal and anterior cingulate, which we saw were important in our fight-or-flight reactions. The activation of our hippocampus accounts for dream memories.

During REM sleep, spinal paralysis sets in so we cannot act out our dreams (which would be a hazard to the dreamer). Our eyes and breathing muscles are left unaffected. Sensations from the outside world are cut off.

Although the purpose of REM sleep and dreaming remains a nearly complete enigma, we do know that REM sleep is a vital function. If we are deprived of REM sleep, we die more quickly than if we stop eating. REM sleep may be necessary for energy regulation: animals deprived of REM sleep fall into a severe energy imbalance before they die. A relationship between REM sleep and memory is clear, but its nature remains uncertain and controversial.

The brain regions turned *off* in REM sleep are as important as those that are turned on. One off area is the dorso-lateral prefrontal cortex. A mouthful to say even for a hardened practitioner, the dorso-lateral prefrontal cortex is central to our capacity for logical problem-solving and "executive" or planning abilities. It organizes information, thoughts, and emotions. It has a role in delaying gratification. Turning this area off is most likely why dreams seem real at the time and we lack the insight that we are, in fact, dreaming. It also helps explain why dreams often seem so out of context and

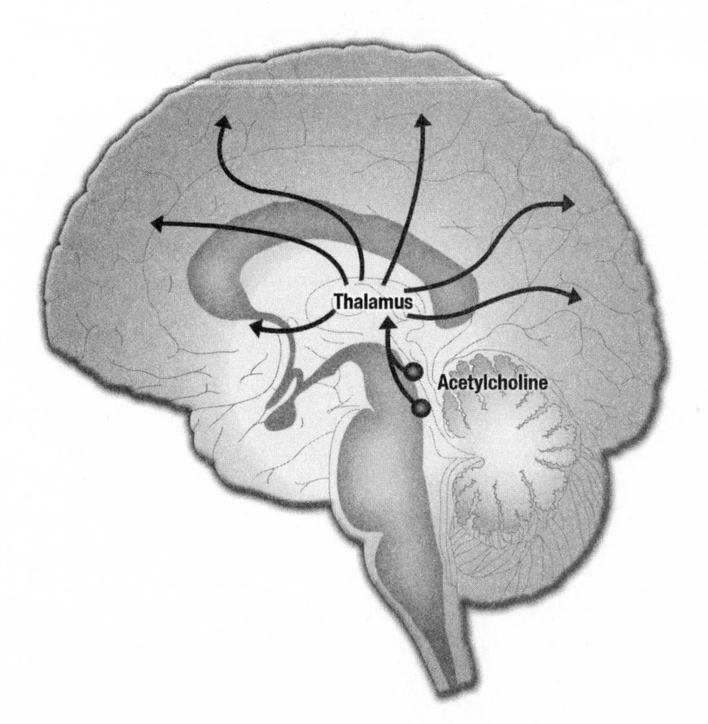

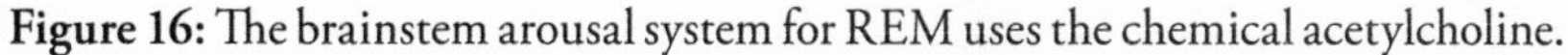
Figure 16: The brainstem arousal system for REM uses the chemical acetylcholine.

disorienting. Time, places, and people change abruptly, in bizarre ways. Hobson writes that dreaming is delirium: "You are as crazy as a loon when you are in the grips of a dream." We act without volition. The madness of our dreams just happens.

Saint Augustine saw dreams differently. In his *Confessions* he wrote: "Surely reason does not shut down as the eyes close." Well, we've seen through scans that the reasoning part of our brain does, indeed, shut down when we dream, but maybe Saint Augustine was on to something that has only recently become clear.

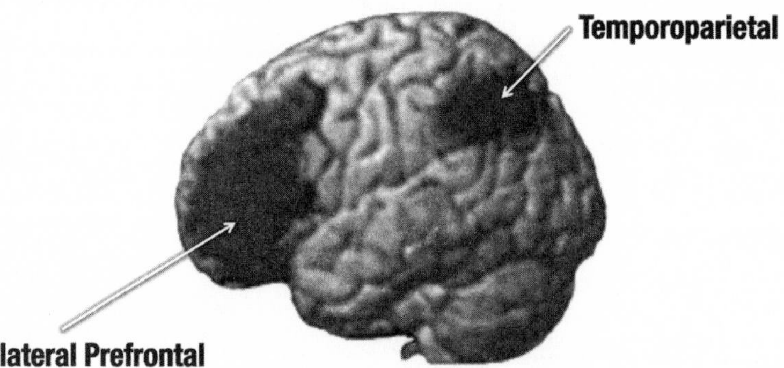

Figure 17: A combined 207 PET scans showing the dorso-lateral prefrontal and temporoparietal regions are turned off (use less energy) during REM consciousness. These findings help explain why dreams are so bizarre, why lucid dreams and NDE feel real, and why REM causes out-of-body experiences.

Lucid Dreaming

If the dorso-lateral brain, instead of being turned off, were active in REM sleep, then we might realize we are dreaming as our dream unfolds. This is called lucid dreaming, a hybrid conscious state that may be very close to the near-death experience.

Lucid dreaming mixes the two conscious states of REM sleep and wakefulness, but only in 3 percent of dreams do people enter this borderland of consciousness. We could think of lucid dreaming as being awake while we dream. The dream feels as real as the world we know when we are fully awake; the experience is balanced precariously between reason and the surreality of dreams.

The lucid dreamer typically enters the borderland from the REM stage of sleep, but sometimes the dreamer can go from awake directly into a lucid dream state, with little sense of a transition.

An anomaly often makes the lucid dreamer suspect that he or she is dreaming: a clock radio that can't be turned off because its buttons are screwy, a book's words that blur, fade, or weirdly change.

We don't understand what brain process distinguishes between the "real" and the imaginary. What is clear is that this process is often scrambled in the borderlands of REM consciousness.

Lucid dreaming is not just a mystifying trait; it has been a spiritual tool in Tibetan Buddhism for hundreds of years. It is, perhaps, important to other spiritual traditions as well. We can learn to make our dreams lucid, and our skills improve with practice. Several techniques have been devised. Most rely on mental exercises or clues that allow us to recognize when we're dreaming.

The experienced lucid dreamer can feel a wide range of emotions, including fear, spiritual ecstasy, and sexual bliss. Sex is a common theme in lucid dreams, and lucid dreams can even culminate in orgasm.

Memories of lucid dreams are more complete than the amnesia that usually follows run-of-the-mill non-lucid dreams: the lucid dream remains vivid even after the dreamer awakes. Look at this account of a lucid dream and see how alike it is to some of the near-death experiences we have examined.

> I realized I was dreaming. I raised my arms and began to rise (actually, I was being lifted). I rose through black sky that blended to indigo, to deep purple, to lavender, to white, then to very bright light. All the time I was being lifted there was the most beautiful

> music I have ever heard. It seemed like voices rather than instruments. There are no words to describe the JOY I felt. I was very gently lowered back to earth. I had the feeling that I had come to a turning point in my life and I had chosen the right path. The dream, the joy I experienced, was kind of a reward, or so I felt. It was a long, slow slide back to wakefulness with the music echoing in my ears. The euphoria lasted several days; the memory, forever.

The feeling of a journey, euphoria, and joy, reaching a turning point, and returning to earth all mirror the near-death experience, as well as the fact that the memory of the experience was burned so deeply into the dreamer's psyche.

A woman wrote to me describing how her near-death experience seemed to be a trigger for a kind of lucid dreaming, where REM sleep and wakefulness mix. It proved to be a fascinating case of the relationship between these borderland dream states and the near-death experience.

Anne contacted me after reading about my research in a magazine. She had been in a car accident fourteen years earlier and had had an experience, which, she said, wasn't spiritual, although it did change her in some ways.

"There was no tunnel or bright light," she wrote. "But I had the out-of-body experience, the life flashing before my eyes (though it was mostly things that I HADN'T done in my life yet), and a sense of calm, happiness, and warmth. Since then, I have no fear of death. Yet, I'm still not convinced there is an afterlife in the traditional sense. I still consider myself an agnostic."

Anne was prompted to write because of "weird and complicated sleep issues" that started after her accident. Nothing to do with spirituality at all.

She wrote: "I feel myself falling back into my body occasionally during dreams. I also have the experience where my dream feels like it continues when I am awake and I also feel myself dreaming before I am fully asleep at night. It's almost like there is a lag between switching over from awake to asleep and vice versa."

Perhaps we should give Saint Augustine his due. He was struck by how real dreams are, and he might have had lucid dreams himself. He struggled with the vividness of dreams so close to reality he felt it necessary to distinguish his moral behavior in dreams and waking.

Because the lucid dreamer is in REM consciousness, her senses are cut off: any sensation she experiences comes from her dream world. She is unable to feel her sheets and blankets or hear the ticking of the bedside clock, but she has some capacity to deliberately control events in the unfolding dream. Lucid dreams are an unstable conscious state, lasting less than two minutes on average. Strong emotions or sensations like pain can bring them to an abrupt halt.

How do we know that someone is really in a hybrid conscious state and not simply awake with an active imagination? Some tangible piece of evidence needs to come back from the dream world to prove that someone was actually in a lucid dream.

No one has studied lucid dreams more intently than Dr. Stephen LaBerge, who gave us the earlier description, and his research team at Stanford University. LaBerge hooked up a number of subjects trained in lucid dreams (including himself) to medical instruments and recorded their brain waves, eye movements, breathing, and muscle tone. Using agreed upon signals, the subjects indicated with eye movements or breathing patterns, in a kind of Morse code, when they realized they were dreaming. Their hands and feet were immobile, weakened by REM sleep. These physiological recordings confirmed that the person was in REM sleep but, remarkably, able to communicate.

There are recent indications that the more lucid the dream, the more active the dorso-lateral prefrontal brain. Recognizing the incongruence of stepping off a building roof and floating in air could signify the dorso-lateral region turning back on while other fragments of REM consciousness are running along. Since lucid dreaming can be learned, turning on the dorso-lateral brain seems to be at least somewhat under our control.

Recently, sophisticated brain-wave recordings have shown that lucid dreaming is a conscious state between REM and waking. When shifting from non-lucid to lucid REM consciousness, the resonant energy between the thalamus and cortex increases and resembles wakefulness. This is the same rhythm, with a tempo of forty times a second, that may bind sensations and thoughts from distant parts of the brain into the wholeness of conscious perception.

REM Consciousness Borderlands

We are asleep, then our eyes begin to move rapidly beneath our eyelids. We actually cross over into another kind of consciousness. No amount of scientific investigation makes that transformation anything less than a wonder of nature. The components of REM consciousness—paralysis, visual activation, and dreaming—engage different brain regions. Each part of the REM system can be functioning independently when our consciousness shifts between REM sleep and waking. REM sleep can fragment, leaving some of its brain regions turned on while others turn off.

This usually happens during transitions between wakefulness and REM, when the colossal chemical systems of nor-Adrenaline and acetylcholine are shifting us from one conscious state to another like swiftly moving ocean tides. A person can get stuck, briefly, in the borderlands between REM sleep and wakefulness. The borderland is an unstable conscious state—within seconds or minutes it reverts to the more stable states of REM or waking.

Although the borderland experience may be strange along with other researchers, we have found waking REM images in the transition between waking and sleep in some people. A kind of dreaming as you are falling asleep. These images typically go unrecognized. Bold or faded dream images can come to us as we slip into sleep or awaken from REM sleep. Often these brightly colored hallucinations are complex and completely formed animations of people, animals, and things. Auditory hallucinations can occasionally appear as voices or music faintly heard or as loud crashes.

The paralysis of REM sleep takes two forms. It usually hits immediately after waking up from REM sleep. REM consciousness continues while the rest of the brain has awakened, and people find themselves lying awake yet unable to move anything except their eyes. They have the feeling that something or someone is sitting on their chest. Terror often sets in; REM paralysis can feel like dying. There can also be the sense that someone or something else is in the room. The paralysis lasts seconds (which can feel like hours), until, with an extraordinary effort, the person moves a tiny muscle to break the spell and strength returns. Many people have visual hallucinations accompanying their paralysis.

Matt, one of my research subjects, is a twenty-seven-year-old narcoleptic who has frequent REM paralysis along with sensed presence. It started when he was in college, living in the dorms, and these incidents are often terrifying for him. He describes his borderland between sleep and wakefulness as follows:

> I have a feeling of hovering over myself. I look around but I can't move anything except my eyeballs. I see things; I hallucinate. In one particular instance I remember there was a basketball on the shelf of my room. It turned into the head of a rotting corpse. It started making noises. I saw a can bouncing up and down. Sometimes it's like there's a hammer beating next to my ear.

> When I go to sleep, my brain knows what's about to happen. I enter this weird in-between state. I see a presence but it's shadowy. It walks across the room and leans down and says things in my ear. I don't know why I don't see good things. The presences I feel and the hallucinations I see almost always have a hostile, aggressive edge. Sometimes the presence will come and lay down on the bed next to me and whisper things in my ear. I try to scream but I can't move. One night this happened five times in a row, each time when I drifted off into sleep in the course of half an hour.
>
> When I'm comfortable, when I have my girlfriend sleeping with me, it doesn't happen. When there's someone in the house with me, it's generally better. It usually happens when I'm alone. I'm thinking that maybe I should get a dog.
>
> My narcolepsy makes it incredibly easy for me to fall asleep. It's not the kind of narcolepsy you see in Hollywood movies, where I fall asleep head first in my soup. It's heavy, heavy exhaustion after I eat. I can sleep 12 hours, and then feel as though I can fall back to sleep an hour later.

Sleep paralysis happens to about 6 percent of people. It can be brought on with stress, fatigue, and sleep deprivation. It runs in families and is found throughout the world. In Japan, it is known as *kanashibari*, meaning "to be bound in metal." In Newfoundland folklore about the experience, the "Old Hag" leaves her body to sit on the chest, terrify her victims, and crush them.

Nobody knows how many thousands of people have memories of being abducted by aliens, finding themselves on board spaceships, undergoing invasive medical procedures that leave no mark, meeting non-earthly beings that convey important messages to humanity or spiritual revelations, or having extraterrestrial sexual encounters. All these can be attributed to lucid dreaming REM intrusion and REM paralysis. A research team in London headed by Christopher French, who studies the fallibility of memory, investigated a group of abductees. He found they had a high incidence of sleep paralysis and were psychologically prone to interpret their experiences as alien abduction.

REM paralysis takes another unusual form when someone is fully awake: sudden leg or knee buckling after laughing or with strong emotions, surprise, or being startled. Matt has these symptoms as well. "Anything that makes [you] suddenly laugh can trigger it," he says. "A sudden change in mood. I have a helpless feeling of cataplexy, a sudden involuntary muscle loss." This feeling of cataplexy is similar to the feeling of weakness that washes over us when we are helplessly tickled. This is also similar to a kitten becoming limp when the mother cat picks it up by the scruff of the neck.

The REM Switch at a Moment of Crisis

How does someone get caught in the borderland between REM sleep and waking consciousness? A switch in the brainstem tilts us between these two states. The switch mechanism lies near the locus coeruleus. It's made up of different components that are small and indistinct even under the microscope unless highly specialized techniques are used. Some of these components tilt consciousness to REM, and others tilt us awake. The switch operates in an all-or-none, winner-take-all fashion most of the time and almost always smoothly shifts the brain between REM sleep and waking. The switch is incredibly powerful—just a few thousand neurons hold sway over the other billions, including those in the dorso-lateral prefrontal region.

Patients with narcolepsy have a chemical deficiency that causes their REM switch to tilt rapidly and frequently to REM consciousness. Sleep attacks, REM intrusion, REM paralysis, and dream imagery plague narcoleptics like Matt; their fundamental abnormality is their inability to control the borders between the brain's three conscious states.

A critical component of the REM switch has recently come to light through an elegant investigation by neurologist Cliff Saper's team at Harvard. Saper's research helps reveal the relationship between blood pressure, fainting, near-death experiences, and REM sleep. He's found that tucked away near the center of the brainstem is a portion of the REM switch called the vlPAG. When it activates, consciousness tilts toward waking and away from REM. The narcoleptics' vlPAG is sluggish.

What the vlPAG does (or doesn't do) when our blood pressure drops is fascinating. It activates when we experience pain, low amounts of oxygen in our blood, or blood loss that causes low blood pressure. That makes sense. We should be wide awake, full of nor-Adrenaline from an energized locus coeruleus, when our brain's blood supply is threatened. We need to be in full fight-or-flight mode.

But something profoundly changes when the pain is strong and inescapable or blood pressure is severely low, as in fainting and cardiac arrest. In the case of severe blood loss, the vlPAG nerve cells cause the sympathetic response (adrenaline and nor-Adrenaline) to pull back, bringing the acetylcholine parasympathetic system to the fore. The heart, instead of beating rapidly to maintain blood pressure, slows, and blood pressure falls even further.

Why on earth would this happen?

During severe pain or blood loss, as the vlPAG retracts adrenaline and noradrenaline and the heart slows and blood pressure drops, the person (or animal), once agitated and visually scanning the surroundings, becomes very quiet and inattentive. The person disengages from his or her surroundings. Remaining quiet and still when the injury is severe and inescapable may be an effective survival strategy—playing possum, playing dead, ceasing to struggle. Whatever its advantage, it is effective enough to have become hardwired into our brainstem. Our locus coeruleus shifts from high discharge rates to a slow and quiet pulse. And what brings it to this slow beat?

It is as if we are designed to pause. Not only do we have a panic button, but a calm button, too.

Normally when the vlPAG pulls back, REM sleep immediately follows. It is no coincidence that the only time the locus coeruleus is completely silent is during REM consciousness. It is extremely significant that switching to REM is *the* most powerful brake on the locus coeruleus.

The REM switch and locus coeruleus connection reflects that successful behaviors like fight-or-flight or lying quietly *must* be tightly coupled to the right conscious state, and gives us strong clues to how our consciousness might tilt to REM in crisis. Ultimately these facts alone suggest that consciousness is directly connected to visceral responses during fight-or-flight.

As *Homo erectus*'s blood pressure failed and he lay limp in the big cat's jaws, he personally could have been well served by escaping into REM consciousness as he died. When David Livingstone, the famous Scottish missionary and explorer, found himself in a lion's jaws, the shock, he wrote, "caused a sense of dreaminess in which there was no sense of pain or feeling of terror, though [I was] quite conscious of all that was happening." Livingstone attributed his reaction to the "merciful provision by our benevolent Creator for lessening the pain of death." The troop of the doomed *erectus* certainly benefited when the cat broke off chasing them since its meal was no longer struggling to escape.

A New Borderland of REM Consciousness

Until recently it was assumed that near-death experiences arose strictly in the cerebral cortex. After all, the cortex is the most highly developed brain region, the seat of reason, language, and all our other distinctly human qualities. It was natural to look for spiritual experience in that part of the brain that most sets us apart from other animals.

Although their link to REM consciousness seemed clear when I first looked at near-death experiences from a neurological perspective, demonstrating the connection was another matter. I knew REM caused paralysis, understood there was a consciousness switch, and recognized how these might be connected to the fight-or-flight chemical systems in the body and brain. But how to show this link? There's the rub. It was a scientifically daunting task.

Although many people have their near-death experience in a medical setting, it's nearly impossible to predict who will have one or when they will be near death. So how can you study it?

My research team and I came up with what turned out to be a simple plan. If REM consciousness sparks the near-death experience, and since not everyone has a near-death experience in crisis, maybe the arousal brain in people who have these special experiences *predisposes* them to blend REM and wakeful consciousness. Some people, because of the way their brains work, might be susceptible to blending REM and waking consciousness, not only during the crisis of being near death but at other times as well.

With the help of an Internet community committed to providing information and support to people who have had near-death experiences, we located fifty-five subjects who willingly participated in our research. As you'll recall, in our study each person believed at the time of the near-death that his or her life was in immediate danger, and each person had to score at least seven on the Greyson near-death scale.

What we found startled us.

We asked the subjects about their lifetime sleep experiences, using exactly the same questions for everyone. Specifically, we wanted to know about the borderland between wakefulness and sleep. We asked if during this transition they had ever experienced visions, sounds, or paralysis, in other words REM consciousness.

What we found was that the brain switch linking waking and REM consciousness was *different* in people who have had a near-death

Table 5: REM intruding into wakefulness for fifty-five near-death subjects compared to a control group of equal number, age, and gender.

	NDE	Comparison Group
REM visions while awake		
"Just before falling asleep or just after awakening, have you ever seen things, objects, or people that others cannot see?"	42%	7% **
REM sounds while awake		
"Just before falling asleep or just after awakening, have you ever heard sounds, music, or voices that other people cannot hear?"	36%	7% *
REM paralysis while awake		
"Have you ever awakened and found that you were unable to move or felt paralyzed?"	46%	13% *
"Have you ever had sudden muscle weakness in your legs or knee buckling?"	7%	0%
Any REM intruding into wakefulness		
One or more types	60%	24% *
One type	16%	20%
Two or more types	44%	4% **

** How can we tell if REM intrusion and near-death occurring together is not simply coincidence? Using statistical calculations, here we find the likelihood of REM intrusion happening by chance in near-death subjects to be less than one in ten thousand. For a rough idea of the odds, pick a number from one to six and throw the dice. We expect by chance that number to come up five times in a row once in a series of about ten thousand tosses. It happens but not very often.

* These findings happen from chance alone less than one in a thousand times (four consecutive throws of the dice showing our selected number).

experience. Instead of passing directly between the REM state and wakefulness, the brain switch in those people was two-and-one-half times more likely to blend the two states. Across the board, people who'd had near-death experiences had *more* REM intrusion—be it REM's dreamlike visual imagery, auditory hallucinations, or paralysis. It was fifty-fifty if the person first experienced near-death or REM intrusion.

It is also significant that those people with visual or auditory REM intrusion had higher Greyson scores during their near-death experience, suggesting that their experience may have been more vibrant because of their REM tendency.

Our near-death experience study subjects clearly had a different arousal brain from most people's. This difference in the arousal brain is likely close to the REM switch itself. It is not only one part of REM consciousness that mixes with waking consciousness in these people—*all* parts do (paralysis and all kinds of hallucinations), and all the parts neurologically converge at the REM switch.

Maybe the vlPAG is more likely to trigger REM consciousness in these individuals, when it pulls back the sympathetic nervous system in the face of inescapable pain or blood loss when they are having a brush with death.

The REM switch—and, within it, the vlPAG—gives us a solid framework linking REM consciousness to near-death experience. Our study strongly showed that people who have had a near-death experience possess an arousal system predisposed to blending REM and waking consciousness.

REM Intrusion Is Not a Fluke

Some people have argued the dreaming of REM sleep can't possibly explain the undeniable feeling of ultimate reality and the emotional power of a near-death experience. These feelings of transcendence are just not like dreams—period.

But do we know what dreaming is? Can we say we really know what transcendence is? Meanwhile, we know what is happening during REM consciousness and how it happens.

Foremost, REM intrusion happens both when we're healthy and when we're sick. This is significant. Sleep paralysis and dream imagery in the transition between REM and waking is common in otherwise healthy people. In our study alone, REM intrusion happened to roughly a quarter of the people without a near-death experience that we randomly selected. More people have the potential for having a near-death experience than is often realized. This may be one way we can be spiritually gifted. And yet REM intrusion can reliably be brought about in the laboratory in an otherwise healthy person awakened from deep sleep. After resuming sleep, the person will reenter REM within minutes, and if again awakened, he or she is likely to have REM consciousness while awake. Not just a little REM consciousness, but paralysis with vivid dream imagery that can contain sight, sound, and touch.

But then, some people get an awful lot of the REM state. Narcoleptics, of course, have frequent REM intrusion; their REM switch tilts to the REM-on position at the slightest provocation. They are often in REM almost immediately upon falling asleep; opportunities for their blending of REM and waking consciousness abound. Sometimes a brainstem stroke will upset our body's natural balance and release the REM system from its normal constraints. Brainstem stroke victims have reported dream imagery while awake: a tunnel with a "golden gate" at one end, angels, and the sensation of levitation. Delirium tremens (brought on by alcohol withdrawal) causes an adrenaline surge and hallucinations (the famous "pink elephant")—symptoms of unrestrained REM consciousness. Parkinson's disease is a degeneration of nerve cells in the brainstem producing slow muscle movements, tremors, and stiffness. In its later stages, Parkinson's generates vivid hallucinations that come in part from REM intrusion; the disorder blurs the boundaries between conscious states and resembles narcolepsy.

There can be so much REM intrusion in narcolepsy, delirium tremens, and Parkinson's that it is difficult to distinguish between REM and waking consciousness in people afflicted with these conditions by looking at their brain waves and other physiological recordings.

Guillain-Barré syndrome—another neurological disorder that causes REM intrusion—is an uncommon attack on nerves outside the brain, launched by the body's immune system. The most severely affected nerves go limp, along with breathing muscles and the autonomic nerves supplying vital organs like the heart.

Guillain-Barré has many causes: it can be triggered by H1N1 influenza or its vaccines. Most people recover in several weeks or months. Early in the illness, there can be severe damage to the autonomic nerves bringing information to and from the heart and blood vessels. The heartbeat becomes erratic—too fast or slow. Blood vessels constrict or dilate, cutting down blood flow or diverting it from where it's needed: the brain and heart, for instance. This can cause adrenaline surges and sometimes a fatal fluctuation in blood pressure.

In Paris, neurologist and sleep specialist Dr. Isabelle Arnulf and her research team studied more than a hundred patients with Guillain-Barré. She observed that the disease produces sleep that is fragmented and unstable. In her patients, consciousness frequently and abruptly shifted between waking, REM sleep, and non-REM sleep. REM began abnormally at the onset of sleep, and, often, the boundary between REM sleep and wakefulness was blurred.

Arnulf found that the autonomic storms of the disease induced severely altered conscious states and visual hallucinations; as soon as her patients closed their eyes they saw goblins and other apparitions. During sleep, her patients had strange and elaborate dreams of flying or leaving their bodies. These experiences were intensely remembered even months later.

How in the world could nerves in the body below the neck that are affected by Guillain-Barré cause such vibrant instances of REM consciousness? What this means is that something within the heart, lungs, or gut can trigger profound changes in consciousness—REM consciousness in particular.

Here the cardiologist and neurologist part ways in understanding how the body and brain are inexorably and interdependently connected. This is the long-hidden crux of the mystery of near-death experiences.

REM Intrusion and the Heart's Nerve

No matter how much human biology someone may know, if that person can't see how the heart and consciousness are connected, then it will be impossible for him or her to fully understand how REM consciousness and near-death experiences are linked. The spiritual doorway has a lot to do with your guts.

Nerves that run to the heart are an obvious place to look for clues about the near-death experience, because these experiences are most often caused by conditions that disturb the heart in some way. Furthermore, we know these nerves can, by themselves, produce dreams.

In order for the brain to control its blood supply (blood pressure) second by second, the brainstem receives information as quickly as the nerves can convey it from all parts of the body. The brain, as we've seen, also ensures that we are in the proper consciousness state to attend to basic metabolic needs like oxygen and glucose for energy, as well as to successfully execute fight-or-flight or to disengage when the blood pressure precipitously drops. Consequently, our brain tightly locks in our vegetative functions with its consciousness state. It is a bit like an airport's control tower directing traffic, using information from radar and radio contact with incoming and outgoing aircraft. The controller must be properly *attentive* to safely orchestrate flow.

During times near death, when blood pressure is low, nerves frantically transmit information to the brainstem from a variety of sensors in the heart, blood vessels, and lungs so the brain can ensure its supply of blood, bringing oxygen and glucose. Most of this information travels by way of the vagus nerve.

The vagus is massive. It is a nerve that is part of the parasympathetic nervous system, and it uses the chemical acetylcholine to transmit its impulses. This is the same chemical the REM system uses in the brainstem. Eighty percent of the vagus conveys information to the brainstem. And it is far and away the heart's dominant controller. It determines heart rate. Fainting occurs when the vagus nerve is overactive, precipitously slowing the heart and lowering the blood pressure. During the autonomic ups and downs of Guillain-Barré, the vagus nerve is sending so many chaotic messages that the brainstem just can't keep up. Could wild vagus activity alone tilt someone into REM consciousness?

It seems so. All mammals have REM consciousness, and if we electrically stimulate the vagus nerve in laboratory rats and cats, a swift and sure tilt to full-blown REM consciousness follows.

Vagus nerve stimulation promotes REM, elicits PGO waves, the "lightning bolts" mentioned earlier, and causes paralysis. The shift is so strong and abrupt that some researchers have called it "reflex REM narcolepsy" or the "narcoleptic reflex."

The story in humans is understandably a bit more complicated: we don't invasively study our patients in the same way we do laboratory animals. We do know that an abrupt tilt into REM consciousness brings on a blending of REM and wakeful consciousness. When the vagus nerve is stimulated for medical purposes in patients, REM appears rapidly with sleep onset, and REM intrudes into non-REM states.

If we follow heart and blood vessel sensors upward with the vagus nerve, they reach the acetylcholine nerves critical for REM consciousness in the brainstem. Near the REM switch, nerve cells coordinating blood pressure and breathing mingle with acetylcholine nerves active in REM consciousness. How these blood pressure, respiration, and REM-active cells live side by side and if, by themselves, they can switch us into REM consciousness remains to be seen. But their proximity and interrelationships are not coincidental.

What is clear is that through its nerves the heart can cause REM consciousness in waking times. The relevance of this to near-death experiences is profound.

All the Way Out of the Body

How could REM intrusion trigger aspects of the near-death experience like feeling as if you have left your body? We saw in an earlier chapter how out-of-body experiences occurred when the temporoparietal junction brain region became disturbed, out of whack in terms of its normal function. We saw that out-of-body experiences could be caused by fainting. Now we're going to go up a notch. Out-of-body experiences are affected by arousal and REM systems. They also have a clearly discernible role in near-death experiences. That role includes contributing to a vivid sense of a transcendent self, an encounter with a higher being or power, and merging with universal consciousness.

The connection between our arousal system and out-of-body experiences makes intuitive as well as scientific sense. Danger provokes a fight-or-flight response, be it a car crash, falling out of a building, or any instance of imminent mortality. The symptoms of posttraumatic stress disorder include chronic fear, unwanted memories, *and* out-of-body experiences.

A neurologist I know was once in charge of an emergency room when he got a call that a seriously injured patient was en route. The neurologist was consumed with anxiety. Standing at the emergency room entrance, anticipating the ambulance, he found himself hovering over his body, looking down. Fortunately his experience was over by the time the patient arrived.

The connection of out-of-body experience to arousal and REM consciousness has been suspected for some time. Narcoleptic patients often have out-of-body experiences, which diminishes when their narcolepsy is treated.

Lucid dreamers are prone to out-of-body experiences, too. When Dr. LaBerge's research team looked at lucid dreaming, verified during brain-wave recordings, almost 10 percent of the dreamers said they had left their bodies.

Let's return to figure 17 that shows which parts of the brain are turned off in REM consciousness. One inactive region is the dorsolateral prefrontal. Another is the temporoparietal. Blanke demonstrated that when this region is disturbed by electrical stimulation, we can expect an out-of-body experience. Maybe something similar happens when the temporoparietal brain shuts down in REM sleep

The following case illustrates this connection between REM consciousness and out-of-body experiences.

Karen was a psychiatry resident physician-in-training at the University of Kentucky. We were introduced by a mutual friend who thought I might be interested in Karen's experiences. We met after work so I could quietly hear her story away from the bustle of the hospital. Karen was initially shy and reserved, but after gentle prompting she told me about an out-of-body experience she'd had fourteen years earlier. Here she tells the story in writing:

I was a twenty-one-year-old newlywed when we moved to the foothills of eastern Kentucky, where my husband was beginning his medical practice. I was raised as a Roman Catholic in a rural French Acadian fishing and farming community in eastern Canada. Although our new town was also in a rural area, the geography and Bible Belt culture were very different from where I grew up. Shortly after we moved, a state trooper next door discharged his gun, under suspicious circumstances, almost killing his wife. This greatly upset me.

That night I was lying anxiously in bed, halfway between being awake and asleep. Suddenly, without warning, I was floating a few feet above my bed, surveying what lay below me. I was confused and uncertain if I was awake or asleep. Looking down, I saw my husband and me sleeping, both of us oblivious to what was happening to me. Although the light was dim, I could easily make out the brightly colored quilt I had made for our bed.

My initial fright soon gave way to curiosity. I wondered if I could float to other parts of the room—and then I did! I looked down at our sleeping bodies from different angles. I was simultaneously frightened and amused. Where else could I go? Maybe I should try to go to the next room? The last thing I remember before waking up the following morning was a strong fear that if I left the comfortable space in which I hovered, I could lose control. I don't recall how I left or returned to my body.

I often remember this experience, but I have spoken about it to very few people. It was just too bizarre—what would people think?

Karen said she didn't see this experience as spiritual. Later I learned that she had had several episodes of sleep paralysis, as had many of her siblings (she came from a large family). She couldn't tell me if her body on the bed was paralyzed at the time of her out-of-body experience; she was too excited by being able to float around the room to notice.

I learned that her father had once had a somewhat similar experience when he lay on a gurney in the emergency room. He didn't notice anything amiss until he began to float above his body, which was prone and still. He was perplexed when he looked down and saw a nurse rushing over to put two paddles on his chest. Next, he remembers a thumping shock and drifting back down to the stretcher. The nurse explained to him that his heart rhythm was "out of whack."

One sleep expert found a simple but clever way to determine if an out-of-body experience happened while awake or in REM sleep. He asked his research subjects who had out-of-body experiences to put some object in an unlikely place before going to sleep and concentrate on the objects in their familiar bedroom while they floated outside their bodies. No one saw the unusually placed objects during their out-of-body experience. This shows that the visual image during out-of-body experiences is not formed while floating around in the room. Instead it suggests, but, of course, does not prove, that the image subjects had while floating was constructed from familiar memory.

There were other ways their out-of-body experiences did not correspond to waking reality. Clocks displayed impossible times or a design that did not correspond to the clock face itself. "I check my alarm clock, and if the bright green LED is not there, then I immediately know that I'm having a sleep disorder experience," said one research subject. And the person sleeping may not be the person floating out-of-body. Another subject reported: "I looked at 'me' sleeping peacefully in the bed while I wandered about. Trouble is the 'me' in the bed was wearing long johns [which] I have never worn."

Many people I've talked to who have had out-of-body experiences have also had REM intrusion with sleep paralysis. Patrick, who had his near-death experience on the ambulance run, had had several episodes of sleep paralysis followed by out-of-body experiences. In

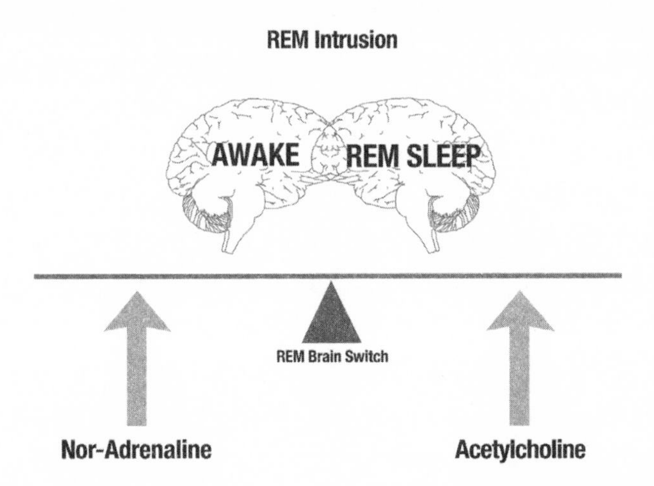

Figure 18: REM and waking consciousness blend when the REM switch is not tilted completely in one direction or another.

one, he fell asleep watching late night TV. Within minutes he awoke, unable to move. He recalled: "I began to levitate. I couldn't move my arms or anything as my body slowly rose." He started out above the couch, at the same level as the TV, but soon he was looking down at the Jay Leno show. This didn't last long and he slowly sank back down to the couch.

Anne, who earlier described her lucid dreams, says she occasionally has out-of-body experiences with REM paralysis.

There are strong indications that out-of-body experiences and REM intrusion are connected. In a large survey of college students, almost 30 percent had had sleep paralysis at least once; some also had out-of-body experiences. This survey was conducted by the same researchers whose results bolstered the tie between REM consciousness and sensed presence during sleep paralysis and dream imagery.

When my friend Jake awoke at 3 A.M. and sensed his mother's presence, he was most likely in REM sleep. I expect he was dreaming of his mother, whom he'd been worried about all week, knowing she was on the verge of death. The moment he awoke was close to, but may not have been exactly, the moment his mother passed. And when he abruptly awoke, I believe his temporoparietal region was still turned off from REM, giving him her sensed presence. Her smell and breath could have come from an important early memory. Despite all this explaining, nothing takes away from Jake the emotional importance of what happened. He felt a connection to his mother that will no doubt always be a part of his life. It would be foolish to think of it as somehow counterfeit. The mysterious systems that enabled it are ancient.

In our research we were surprised to find that out-of-body experiences were as likely to occur during the transition between waking and sleep as during near-death experiences. We also found strong evidence that people who have near-death experiences have a brain arousal system in which the switch linking waking and REM sleep is not only wired for REM intrusion in near-death experience but for out-of-body experience as well.

The Unearthly Light

One of the most startling and consistent features of near-death experiences is unearthly light. Again REM intrusion helps explain how the brain is functioning in this part of the near-death experience.

We have already seen that in many cases of near-death experience the REM switch activates electrical waves that radiate from the brainstem upward, energizing the visual brain. At the same time severely low blood pressure causes the eye's retina to lose peripheral vision, creating a tunnel. Light at the end of the tunnel can come from two sources.

The first is ambient light that comes through our eyelids, which are half-open as the brain struggles to maintain consciousness. The light strikes a retina starved for blood. Light distorted by a failing retina is then conveyed by a visual system to a brain that is on the verge of unconsciousness. By this time, smudges of outside light at the end of the tunnel may be all the brain is capable of seeing.

A second source of light might, of course, be the light of REM consciousness. When outside light does not reach the brain, then the REM system reigns supreme over the visual brain. Light is the core business of the REM system, creating the visual images that play such an important role in our dreams and REM intrusion. I think it's likely that the light of REM consciousness gives near-death experiences their unearthly quality, which people also experience during lucid dreams.

Even if the brain lacks so much blood that it becomes blind, it retains its ability to create dream imagery. A stroke may leave its victim unable to physically see, but that person will still dream in images. REM consciousness functions if outside light is cut off where it first enters the visual cortex. In fact, that region is normally shut down in REM sleep, cutting us off from the outside world; other parts of the visual brain produce the light of dream images.

Bliss Near Death

The reward system that we saw connected to the fight-or-flight response also has a strong bond with the REM brain, and that is our starting point for understanding the rapture of many spiritual experiences.

The acetylcholine nerves instrumental to REM consciousness connect to the brain's reward center. Heroin, cocaine, and alcohol all use this pleasure center. If REM regions are injured in animals during lab experiments, they lose nearly all their interest in food or heroin should they be subjects of substance abuse research.

The connection between the brain's pleasure center and REM consciousness is unmistakable. During REM consciousness, cells within the reward center are vigorously energized. It's as if we are eating delicious food or having sex. Brain images of people sleeping show regions closely connected to the reward center are powerfully active during REM consciousness.

We can't directly connect the pleasures of near-death experiences and fainting to the reward system (no one has placed his or her head in an MRI at exactly the right time). Yet the scientific path is clear. Regardless of what gaps remain in our knowledge, experiencing the bliss of heaven while on earth must have its genesis somewhere in the brain's reward system.

Divine Dreaming

The evidence from several quarters strongly suggests that the heart, so often at the center of peril during near-death experiences, can trigger REM and the brain's dreaming machinery. By doing so, maybe the heart brings us dreams when we need them most.

Still, many people are hesitant to accept the role of REM consciousness in near-death. "My near-death experience can't be a dream, it was so different from any dream I ever had." "My experience felt so real; I remember everything. I hardly remember any of my dreams." I've often heard these objections after I explain the link.

My point is that near-death experiences are not like our nightly dreams because these experiences are *not* dreams in the usual sense. They are more akin to the lucid dreams of the REM-waking borderland. Part of the dreaming brain erupts in a brain already awake. And blending REM with waking consciousness creates experiences that are realistic and memorable.

Many military pilots—who have extensive experience and training with fainting—have reported "dreaming" in the borderlands of consciousness since the earliest days of aviation. The more brain blood flow is interrupted, the more likely it is that the fainter will dream. Estrella Forster and James Whinnery, who have studied hundreds of pilots at their U.S. Air Force–based laboratory, deliberately caused eight volunteers to faint in a human centrifuge. A quarter of the time their fainting was accompanied by "dreams," and the fainters knew they were dreaming. They found these dreams intense and memorable.

The fainting dreams were of two types. One type was characterized by fear and confusion; the other, euphoria and light. The dreams happened during paralysis, which the researches likened to sleep paralysis. Unfortunately, no researcher has yet captured fainting and REM consciousness on physiological recordings (or we have, as yet, been unable to recognize it).

Pilots can experience a "break-off" illusion, so named in aviation medicine because the pilot feels removed from his surroundings and breaks off from reality. In its extreme form, pilots leave their bodies and watch themselves from outside the aircraft. One combat pilot, now a neurologist, told me he found the out-of-body experience he had while flying his fighter jet unbelievably strange. Fortunately, he said, it quickly passed.

Although the full biological purpose of dreaming remains hidden from us, some new ideas are coming to the fore that could be seminal. Allan Hobson has proposed that REM consciousness prepares brain circuits for waking consciousness. REM activity begins early in the womb, and newborns spend much of their time in a REM state. Hobson speculates that REM consciousness could be the beginning of human consciousness, where we first become aware and develop our sense of being the agent of our actions, thoughts, and feelings. REM consciousness could be the origin of the self. A great deal about REM as "protoconsciousness" makes physiological sense, and I am eagerly watching this idea develop.

REM sleep began evolutionarily in the earliest mammals, so it is reasonable to expect that REM consciousness would be important to survival. Dreams often simulate threatening events, and dreaming can be seen as a virtual simulation of danger. We rehearse these events in our minds before confronting them when we're awake. If dreaming helps prepare waking consciousness for life-and-death struggles, then we should not be astonished if the dream mechanism is activated by a hardwired trigger to REM consciousness—a survival response leading us back in to a more primitive consciousness in crisis. So it may not be accidental that brain imaging finds that the regions activated in danger and REM consciousness share the same ancient limbic system.

That dreams and danger share this limbic system explains so much about spiritual experience. These amazing moments in our lives are part of and in concert with our survival instincts; they are deeply embedded within us—contained in our existential core. It's not surprising that our belief systems often feel as if they are matters of life and death—they are!

Scientists have extensively characterized dream narratives, cataloguing how often this or that does or does not happen. Near-death experiences have not received the same degree of scrutiny. Aside from Greyson's scale, we have only individual stories, that can either support or refute the characteristics of dreams. This doesn't get us very far.

Yet there are things we can learn by examining the overlap between near-death experiences and dream, showing that they share the same limbic landscape.

Dreams are stories. But they are stories told not so much with language but with visual images and emotions. Vision and its medium of light are paramount. This also holds true for many near-death experiences.

We don't dream every minute we are in REM sleep. Some of our dreams are short; others, long. Some near-death experiences have a strong narrative quality; others don't. People, familiar and unknown, play important roles in both near-death experiences and dreams. At least half the time in dreams we encounter people we know by name; it's unusual to encounter complete strangers. So it is with near-death experiences. Our research subjects were more likely to meet someone they already knew or a familiar religious figure, rather than a complete stranger. Recognition in dreams can come from facial features, voice, or behavior, though as often as not the dreamer recognizes a person by "just knowing." This is similar to a near-death experience, where often a person or spiritual entity is also "just known" and can, sometimes dramatically, give reassurance or guidance telepathically, without language. Dream characters reveal themselves to us by the feelings they evoke. Most often these feelings are affection and joy. Knowledge gained through feelings is a limbic kind of knowledge. So too is the knowledge gained from near-death experiences.

Pain is rare in both near-death experiences and dreams. The absence of expected pain is often cited as one of the marvels of these experiences. Being cut off from our bodily senses is a cardinal feature of REM consciousness. Fear, joy, and anger are the emotions that predominate in dreams. These are also the emotions commonly expressed in near-death experiences. On the other hand, sadness is uncommon in both dreams and near-death experiences.

Time is not sensed the same way in waking consciousness as it is in dreams or near-death experience. Things can happen instantaneously and people or places shift abruptly. In dreams, this change is bizarre, probably due in measure to the dorso-lateral prefrontal brain being turned off. Near-death experiences can instantly transport people to new and fantastic places.

The Near-Death Trinity

Near-death experiences work in the brain as a confluence of low blood flow to the eyes, fight-or-flight reactions, and the triggering of REM consciousness. Physiologically, that's it.

My notion of REM intrusion during near-death experiences raises little controversy among neuroscientists. The strongest criticism I hear is that we need more data. This is a common call among scientists—and I couldn't agree more. In any case, I don't see REM intrusion as the last word on near-death experience, nor REM consciousness as the last word on spirituality.

Nonetheless, my REM intrusion hypothesis has been recognized as having two principal strengths. First, it offers a comprehensive explanation for the near-death experience based on well-established brain mechanisms. Secondly, the hypothesis is "testable" in scientific terms. It can be rigorously scrutinized. As more data comes in, our understanding of the brain will change, perhaps more radically than we can now imagine. That's fine. This is the beauty and power of science.

Table 6: Contributions to near-death experience briefly encapsulated.

Near-Death Experience Feature	Physiology
Tunnel	Low blood flow to the eye's retina
Light	Ambient light and REM visual activation
Appearing "dead"	REM paralysis
Out-of-body	Temporoparietal REM deactivation
Life review	Memories (hippocampus) from fight-or-flight
Bliss	Reward system
Narrative Quality	REM dreaming and the limbic system

PART THREE

THE OTHER SIDE

8

THE BEAUTY AND TERROR OF ONENESS

DEEP WITHIN THE MYSTIC'S BRAIN

"Is man related to something infinite—that is the central question."

—Carl G. Jung

Studying William James's views on spiritual experience has encouraged me to look more closely at an aspect of near-death experiences that I initially overlooked but that now looms large in my mind. A sizable 42 percent of our research subjects felt "united, one with the world" during their near-death experience. This mystical sense of oneness was considered by James to be the prime inspiration for organized religion and the most important type of spiritual experience.

What can the brain systems that we've been examining tell us about the mystical? I am not convinced that sensing unity comes directly from the REM portion of arousal. The part of our arousal system linked to fear and fight-or-flight is more likely to help us unravel what's happening in the brain during mystical experience. Our near-death experience subjects who felt unity also had stronger limbic feelings of peace, joy, and understanding compared to the rest. The mystical group was also more likely to have reached some sort of uncrossed border (from which "no traveller returns"). They had slightly more REM intrusion than our other subjects, but not enough to be statistically significant.

The role of the REM state in the mystical may be indirect. Once the limbic system is activated by fight-or-flight or the REM switch, the mystical feeling of unity may depend more on the limbic system itself and less on the brainstem activity that sparked it. When Cliff told me about feeling "a major connection to something larger than myself" as his adrenaline surge waned, it reminded me of what a fellow neurologist told me when he heard I was collecting cases of spiritual experiences. Reed's experience, like Cliff's, links the adrenaline surge, fear, and the transcendent experience of oneness.

The Rising Sun

Reed had just completed his training in neurology in Boston, and with two friends he decided to celebrate the event with a camping trip along the Maine coast. They fell into excessive libation their first night out and were hungover the following morning. To clear his head, Reed decided to take a walk along the shore. In his own words:

> The trail turned out more difficult than it looked. I climbed up and down rocks and across fallen trees. I was surprised to find myself light-headed and sweating excessively. My heart raced. I realized I was having an adrenaline surge from the brisk hike, empty stomach, and dehydration from drinking so much scotch the night before. I decided to climb down to the beach and walk along the ocean's edge.
>
> The sun was rising in the east over the water. I gazed at this stunning vista, and my thoughts drifted to the coming changes

in my life. The waves pounded, and the ocean stretched to the horizon. A sudden expansive feeling of the unity of an infinite universe overwhelmed me. I felt myself fused with the vast ocean and sky; but, at the same time, I held on to a distinct separateness. My entire being felt smaller than the molecules in the grains of sand beneath my feet that stretched up and down the beach as far as I could see. The feeling that instantly followed jolted me. I didn't feel euphoria or peace but a strong sense of indifference. The universe was indifferent! It was malignant. It was infinitely vast and I was infinitely inconsequential and insignificant in the face of its incomprehensible scope. I began to feel a terror that came from a crushing sense of helplessness.

This feeling lasted only for a flash. I couldn't bear it. I looked away from the sky, scanning the beach for something to divert me, trying to force my mind away from the oppressive feeling of existential nothingness to more immediate, mundane concerns. My efforts were successful, and the feeling of nullity passed. Returning to camp, I was hit with an occasional wave of melancholy, and I became aware that my mouth was parched. I had not eaten that morning. I felt weak and shaky.

Back at camp, I gave my friends no indication of what had happened. I felt strangely guilty, as though I had something to hide. I went to the cooler, looking for fluids and glucose, and found a bright orange, floating in chunks of ice. I cut off a slice and took a bite. A cold explosion of juicy flavor filled my mouth. What an intense pleasure that orange was—I will remember its taste until the day I die. The orange's sweetness brought with it a burst of energy and euphoria, yet its intensity was nothing like the bottomless void I had experienced on the beach.

Reed's experience on the beach was singular—he had never experienced anything like it before, and he felt nothing like it again. His awareness of the infinite universe was clearly distinct from the mathematical concepts of infinity. It was experiential, a simultaneous fusion of *knowledge* and *feeling*. Fear was a vital trait and must have played a vital role.

While Reed felt fear on the heels of sensing unity, Cliff's unity experience occurred after his fear subsided. In both cases, sensing unity and sensing fear were tightly linked, which suggests they may be linked in the brain as well.

Taking Mystical Measure

Our study was not designed to probe the mystical aspect of near-death experiences, although we did ask our subjects if they felt united or harmonious with the universe. Unity, harmony, and oneness can mean different things to different people. What our near-death experience subjects reported could have been euphoria that came from simply feeling good about life, or a glimpse of an afterlife that activated the brain's reward system.

Although James's criteria help us understand mystical experience, they don't give us a means to scientifically measure it. In his work *Mysticism and Philosophy*, W. T. Stace, a philosopher at Princeton, elaborated the mystical criteria set forth by James and made necessary refinements so the mystical could be reliably identified. Stace wrote this work to answer Bertrand Russell's "Mysticism and Logic," which dismissively asserted that mysticism was nothing more than an intense illusion that imparted no true insight about the universe. The psychologist Ralph Hood fashioned Stace's idea of extrovertive and introvertive mystical experiences into the "M" scale that is used today in scientific studies to detect and quantify mystical experiences.

How could an experience that is by its very nature "beyond words" be reliably explored with questions? Although personal religion and spirituality form a complex and dynamic process, that process can be

Each respondent circles +1, +2, or -1, -2, or ?, depending on how the respondent feels the description applies in each case.

Your Experience	Description
+1:	This description is probably true of the respondent's experience or experiences.
+2:	This description is definitely true of the respondent's experience or experiences.
–1:	This description is probably NOT true of the respondent's experience or experiences.
–2:	This description is definitely NOT true of the respondent's experience or experiences.
?:	The respondent cannot decide.

studied with psychological methods. The MRI allows us to accurately measure subjective experience—what individual brains are doing or not doing as subjects experience fear, pleasure, and joy. And along with many other areas of MRI investigation, we are seeing a new interest in the psychology of religion and the "M" scale.

This scale has been widely applied to many different people and shows that the mystical experiences of American Christians and Iranian Muslims are more alike than different. There are not Judeo-Christian, Islamic, Hindu or Buddhist, ancient Greek, Egyptian, or other prehistoric mystical experiences. Instead, what we see are Judeo-Christian, Islamic, and even Neolithic *interpretations* of the mystical. The universal core of the mystical makes good sense if the experience is hardwired in our brains.

Like James, Stace said the central experience of mysticism is the perception of "oneness." The borders between subject (the person) and object (the world) dissolve. In his book *The Teachings of the Mystics*, Stace quotes the Upanishads, the ancient Hindu text, to describe the mystical experience as "beyond the sense, beyond the understanding,

Table 7: The "M,"or Mystical, scale.

	Your Experience	Description
1	+1 +2 –1 –2 ?	I had an experience which was both timeless and spaceless.
2*	+1 +2 –1 –2 ?	I DID NOT have an experience which was incapable of being expressed in words.
3	+1 +2 –1 –2 ?	I had an experience in which something greater than myself seemed to absorb me.
4	+1 +2 –1 –2 ?	I had an experience in which everything seemed to disappear from my mind until I was conscious only of a void.
5	+1 +2 –1 –2 ?	I experienced profound joy.
6*	+1 +2 –1 –2 ?	I DID NOT have an experience in which I felt myself to be absorbed as one with all things.
7*	+1 +2 –1 –2 ?	I DID NOT experience a perfectly peaceful state.
8*	+1 +2 –1 –2 ?	I DID NOT have an experience in which I felt as if all things were alive.
9*	+1 +2 –1 –2 ?	I DID NOT have an experience which seemed holy to me.
10*	+1 +2 –1 –2 ?	I DID NOT have an experience in which all things seemed to be aware.
11	+1 +2 –1 –2 ?	I had an experience in which I had no sense of time or space.
12	+1 +2 –1 –2 ?	I had an experience in which I realized the oneness of myself with all things.
13	+1 +2 –1 –2 ?	I had an experience in which a new view of reality was revealed to me.
14*	+1 +2 –1 –2 ?	I DID NOT experience anything to be divine.
15*	+1 +2 –1 –2 ?	I DID NOT have an experience in which time and space were nonexistent.
16*	+1 +2 –1 –2 ?	I DID NOT experience anything that I could call ultimate reality.
17	+1 +2 –1 –2 ?	I had an experience in which ultimate reality was revealed to me.
18	+1 +2 –1 –2 ?	I had an experience in which I felt that all was perfection at that time.

19	+1 +2 –1 –2 ?	I had an experience in which I felt everything in the world to be part of the same whole.
20	+1 +2 –1 –2 ?	I had an experience which I knew to be sacred.
21*	+1 +2 –1 –2 ?	I DID NOT have an experience which I was unable to express adequately through language.
22	+1 +2 –1 –2 ?	I had an experience which left me with a feeling of awe.
23	+1 +2 –1 –2 ?	I had an experience that is impossible to communicate.
24*	+1 +2 –1 –2 ?	I DID NOT have an experience in which my own self seemed to merge into something greater.
25*	+1 +2 –1 –2 ?	I DID NOT have an experience which left me with a feeling of wonder.
26*	+1 +2 –1 –2 ?	I DID NOT have an experience in which deeper aspects of reality were revealed to me.
27*	+1 +2 –1 –2 ?	I DID NOT have an experience in which time, place, and distance were meaningless.
28*	+1 +2 –1 –2 ?	I DID NOT have an experience in which I became aware of the unity of all things.
29	+1 +2 –1 –2 ?	I had an experience in which all things seemed to be conscious.
30*	+1 +2 –1 –2 ?	I DID NOT have an experience in which all things seemed to be unified into a single whole.
31	+1 +2 –1 –2 ?	I had an experience in which I felt nothing is ever really dead.
32	+1 +2 –1 –2 ?	I had an experience that cannot be expressed in words.

* Reverse scored.

beyond all expressions . . . It is the pure unitary consciousness, wherein awareness of the world and of the multiplicity is completely obliterated. It is ineffable peace. It is supreme good. It is without second."

The mystical epiphany is generally brief. Its description varies from culture to culture. Words used to describe it include: boundless, ceaseless, bottomless, nothingness, fathomless, infinite, empty, void, barren, abyss, abysmal, and absolute.

The core sense of mystical oneness is expressed in two forms according to Stace. The *extrovertive* mystical experience looks outward to the world through the physical senses and finds unity. Everything that can be heard, seen, felt, touched, or smelled is melded into One. A pervasive sense of unity shines through everything around us. This is what Reed felt.

The *introvertive* mystical experience turns inward, often shuttering out the senses. Thoughts, feelings, sensations, volitions, and memories are transcended into a "pure" consciousness, devoid of empiric content. The sense of being a separate self of thoughts, feelings, sensations, volitions, and memories is lost and transcends into the One. Time and space dissolve.

Perceiving unity is the most important thread running through both the extrovertive and introvertive mystical experiences.

Much of what we know in mainstream Western science about mysticism comes from the ideas of Stace and Hood's "M" scale, which has become the principal tool that science uses to measure mystical experience (and one I now use in my laboratory). This scale would be sterile, however, if we did not have the rich continuum of the men and women from ancient times up to the present day whose mystical experiences give flesh and bones to the "M" scale. As we try to look for what happens in the brain during mystical experience, it's helpful to examine this rich trove of experience.

A Mystical Pioneer

Some of the earliest writings that describe the mystical come from Plotinus, one of antiquity's most influential writers about mysticism. His spiritual experiences are not conveyed in a familiar narrative style, but he is very clear about the nature of mystical experience and said that he had achieved union with the One four times.

As is the case with many people who have explored the meaning and nature of mystical experience, there was nothing dreamy or flaky about Plotinus. He was both an able administrator and a towering intellect, with both feet planted firmly on this earth.

Plotinus was Egyptian by birth. Around the year 232 C.E. he traveled to Alexandria as a young man, to study, eventually making his way to Rome. Later when he became a master teacher himself, Plotinus attracted a number of students, many of whom were women, and came to be held in high esteem by influential Romans, including the Emperor Gallienus. He has served as the authority on Plato from antiquity to modern times, and he interpreted his mystical experiences in Platonic terms—how the divine unity of the One is expressed through the unbroken continuity of multiplicity. His writings deeply influenced early Christianity, Islam, Judaism, and Indian philosophy. Saint Augustine turned to Plotinus as a starting point for understanding his own mystical experiences.

Plotinus wrote that mystical experiences are brief, sudden, and unexpected. They can't be willed, and mystical union with the One can't be reached by logic. While it is happening, it is beyond reason and language: "It is absolutely impossible, nor has it time, to speak; but afterwards that it is able to reason about it."

The absolute primacy of the One gives rise to all things and is touched through direct personal experience where the finite individual self merges with the infinite: "For intimate self-consciousness is a consciousness of something which is many: even the name bears witness to this . . . soul and life depending on him and moving to an unbounded unity by his sizeless unboundedness."

Meister Eckhart

Medieval Christian mystic Meister Eckhart, an ardent student of both Plotinus and Saint Augustine, was another influential early writer on mysticism. His writings on his extrovertive and introvertive mystical experiences paved the way to his excommunication from the Catholic Church and influenced Martin Luther and the Reformation.

Many of the circumstances of Eckhart's birth are uncertain, but we do know that he was born in Germany around 1260. He entered the Dominican Order when he was young and was educated first in Cologne and later Paris, where he was a professor until 1303. He arrived in Strasbourg in 1313. It was here that he delivered his many lectures touching on mystical experiences, not in Latin but in his native German tongue, which was remarkable for the time. Eckhart's lectures also brought him recognition as one of the Church's most powerful minds.

The evangelical Dominican Order valued the ideal of self-discovery, and it is this principle that led in part to the pope's declaration that several of Eckhart's writings were heretical. Eckhart emphasized the individual and spent a great deal of his energy exploring the place for reason in faith, believing that only rational creatures could experience the divine (I will shortly call that into question). It is ironic, then, that Eckhart should be credited with such a profound understanding of the mystical state, strikingly similar to the Eastern mystics. His understanding sprang from his own experience.

The unity experience brought Eckhart into serious conflict with Church authorities. Foremost, unity of a man with God was so intimate that it made the intermediary of the Church unnecessary. This was one of the ideas that made Eckhart a lonely forerunner on the path to the Reformation. "But when a person has a true spiritual experience, he may boldly drop external disciplines, even those to which he is bound by vows, from which even a bishop may not release him." This is the power of mystical experience for Eckhart. No wonder he was so troubling to the Church!

After the Oneness is felt, how is it to be understood? Here Eckhart had a more delicate and dangerous balance to strike than Plotinus, who was a pagan. Eckhart had to reconcile the One (or God) with the Christian trinity.

Meister Eckhart's mystical experiences and his sermons helped seal his fate, but Eckhart's real undoing came from being caught in a web of papal geopolitical forces during an exceptionally tumultuous time.

Eckhart's Thin Thread

In his examination of the mystical, Eckhart made clear an observation that is often overlooked. No matter how complete the merging is of the self with the absolute in the mystical, the self can't be completely and irrevocably dissolved, because after the mystical experience ends we return to ourselves. "God has left her one little point from which to get back to herself," Eckhart wrote. Back to the memory synapses of self.

The experience of introvertive oneness can only be appreciated through the irony of simultaneously being aware of one's own separate identity.

It has been asserted that Eckhart made the point that he did not personally have complete union with God and, therefore, could make no claim that he'd achieved divinity, in hopes of avoiding charges of heresy by Church authorities. It is more likely that Eckhart was honestly reporting a keen observation, and one remarkably consistent with how we understand the brain today. We can't *know* oneness without retaining a speck of separateness: self-detection can shut down while self-recognition remains.

The brain can lose parts of the self, as we saw when we "found" our right foot. Of course the foot was not actually lost: the thalamus had closed the gates on the sensation before the foot reached consciousness. We opened that gate when we willed our right foot to the forefront of consciousness.

Slamming shut the thalamic gates could be important in the mystical loss of self. Memories about self may not be retrieved in the moment of mystical union, but, after the fact, the memory of the experience shines bright.

If the brain is laying down memory, can we say the self is lost?

In mystical experience, we could say that the awareness of awareness itself is heightened while self-sensation is cut off, in the same way we are cut off from the outside and inside worlds during REM sleep.

Is closing the thalamic gate, so self-detection does not reach consciousness, a way to provoke introvertive mystical experiences? This is a tenable idea but difficult to scientifically test, unless perhaps we could visualize thalamic activity at the moment self is lost. More data please!

Loss of self does not necessarily produce the mystical. Cotard's syndrome patients lose the pronoun "I" but don't experience union with the universe. While denying their own existence, they retain a strong sense of separateness from the world around them. Cotard's is a psychosis—not a realization of transcendental truth.

Important aspects of the neurological self are within the temporoparietal brain. When that area is inactive, it leads to out-of-body experiences. An inactive temporoparietal junction is also found in people who are disconnected from their limbs, convinced that limb movements are outside their conscious control. If the temporoparietal junction becomes damaged, it heightens the personal trait of self-transcendence and the ability to see oneself as an integral part of the whole universe. The temporoparietal junction may also be central to the ability to see ourselves as consistent, carried through the past, present, and future.

It seems possible that shutting down the temporoparietal brain could contribute to the loss of self in a mystical experience, especially during the "pure consciousness" of Stace's introspective mystical experience. The temporoparietal also shuts down in REM consciousness. That could not only account for out-of-body experiences, but also for the mystical during near-death experiences. Perhaps the connection between mystical experiences and REM consciousness is more direct than I first thought.

The self is contained within consciousness. From the neurologist's perspective, it is difficult to imagine—no matter how powerful the sense of a complete loss of self in mystical union is—that no vestige of the self remains while consciousness is present and autobiographical memories are created.

I believe that Eckhart's paradox—developed centuries ago—speaks to what neuroscience today knows about the brain.

The Jeweled Castle

To understand how we can apply the "M" scale and detect the ineffable mystical state, we turn to Saint Teresa of Avila, perhaps the most famous of all Christian mystics.

Saint Teresa was a Spanish nun of the Carmelite Order in the sixteenth century. She was exceptionally pious and her writings have great clarity and beauty (she was the first woman elevated to doctor of the Church, by Pope Paul VI in 1970). While Eckhart was ostracized from the Church for his mystical sermons, Saint Teresa was instructed by her superiors to write a book about her mystical experiences during prayer, which was intended to be an instruction manual for her fellow nuns. As a consequence, we see her interpreting her experiences of union in the conventional theological framework of her time.

Her spiritual experiences are scattered throughout her works, but some of her most lyrical descriptions are concentrated in her book *Interior Castle*, which recounts her vision of a beautiful diamond or clear crystal globe that is shaped like a castle and represents the soul. Inside the castle are seven mansions, the innermost containing the "King." This mansion is a cloud of the greatest brightness where a

"Spiritual Marriage" with God is as if "the ends of two wax candles were joined so that the light they give is one." Saint Teresa believed that joining of lights was within the reach of all her "daughters."

It is, of course, impossible to know precisely what was going through Saint Teresa's brain during her mystical experiences. But two researchers thought they could understand the way the brain functioned during mystical experience by studying the MRIs of her spiritual descendents. Unfortunately, their work, published first in 2006 and then expanded on in the 2007 book *The Spiritual Brain,* was flawed from the beginning.

The Nuns Study

Montreal radiologist Mario Beauregard and neuropsychologist Vincent Paquette studied the MRI brain scans of Carmelite nuns during experiences when, the researchers asserted, the nuns were "in a state of union with God." For a five-minute period, the nuns were asked to close their eyes and "relive . . . the most intense mystical experience ever felt in their lives as a member of the Carmelite Order." During this time, their brains were scanned to see if certain regions indicated increased cell activity.

There were several problems with this experiment, highlighting the hurdles that confront us as we try to understand the mystical brain.

At the experiment's onset, the nuns, echoing their patron saint, told the researchers that "God can't be summoned at will." In fact, it would have violated one of their basic beliefs if, on command, they had had "union with God" while they were in an MRI. Saint Teresa had clearly written: "You must never beseech or desire Him to lead you along [the] road" to mystical experience because it "shows a lack of humility." As it turned out, by their own report, during the experiment the nuns did not achieve any of the core features of a mystical experience, including the loss of time, space, and feelings of unity.

Still, I was excited when I first looked at these scans. The excitement, however, soon gave way to disappointment. Not one but many brain areas activated during the scans, leading the investigators to conclude: "Mystical experiences are mediated by several brain regions and systems." But what systems? There were a lot of dots lighting up in the brain, but the connections didn't add up to anything that I found meaningful.

Not surprising, all the subjects reported that the experiences themselves and the memory of them recalled during the MRI scan were very different. Visualizing the beauty of the *Mona Lisa* from memory is not the same thing as seeing it at the Louvre. The two experiences activate different brain regions.

I think it's safe to say that what was measured in the nuns was not a mystical experience. Rather, it was a spiritual experience based on memory, emotion, meditation, or some combination of these and perhaps other types of experiences.

The Carmelite study proves that it doesn't really make sense to ask someone to stick her head in an MRI scanner and produce a mystical experience on command. But does this mean that mystical experiences are immune to scientific scrutiny? Not at all. There are other ways to study the mystical and generate valuable data.

Imaging the living human brain in the grips of a spiritual experience is an alluring possibility, but we have to be careful. As already mentioned, MRI scans chart changes in blood flow. Not only does a great deal of critical brain cell activity go undetected, but crucial cell clusters could be too small for the MRI to detect.

A more important and subtle weakness becomes apparent when we study the writings of the mystics. A cardinal feature of introspective mystical experiences is their lack of empirical content. This suggests that critical brain regions are shut down, inhibited from action.

We have seen on PET scans that REM consciousness shuts down the dorso-lateral prefrontal region, which probably creates the surreality of dreams. Similarly, disrupting the temporoparietal junction disturbs the self and may be important in mystical experience. Such inhibition may have little effect on blood flow and therefore may go undetected by the MRI. And even if reduced brain cell activity indirectly reduced blood flow, neuroscientists often overlook and under-interpret this finding.

Mystics usually report that their experiences are brief. How brief? It's hard to say, especially because the mystical alters the perception of time; a feeling of timelessness prevails. Saint Teresa believed the mystical was sudden, a bolt from the blue: "We might liken the action to a flash of lightning." This flash is a technical problem for MRI: it would be washed out when the scan is recorded over five minutes. The problem is compounded when the signals of fifteen nuns are averaged; it's especially true if only one or two nuns had the flash. What if mystical experience depended on a sequence of brain regions flashing on and off in rapid succession? That fact would also be lost in MRI. In short, not knowing the precise moment of a subjective experience greatly hampers MRI detection.

In the future, the ability of the MRI to detect mystical and other types of spiritual experiences will improve by combining it with other techniques, such as recording the brain's electrical or magnetic activity. But even if the MRI pictures realistically illustrated increased brain activity during a spiritual experience, this by no means proves that the active brain areas are essential to that experience. The evidence would be much stronger if *low* activity in these areas prevented spiritual experiences or interrupted them as they occurred. Just the opposite would hold true if the key mystical ingredient were low brain activity.

Let's assume for the moment that we could reliably identify brain regions activated or inhibited during mystical experience. This leaves us with the question of how these presumably limbic regions get activated or inhibited. We must also remember that many different systems can be housed in the same region of the brain. Using MRI to understand how the brain processes during spiritual experience is roughly akin to trying to grasp the political process in Washington, D.C., by studying which street intersections carry the most traffic.

We do have available to us the means to define brain systems of mystical experience with molecular precision. But it is an approach that takes personal courage and a different kind of mystic, one who is committed to traversing the spiritual doorway in the brain by whatever mean possible to investigate this most elusive and profound of experiences.

Are You Experienced?

For centuries mystics have chosen secluded and isolated settings to concentrate on consciousness, turning off both the outside world and inner sensations. In 1954, a new mystic emerged, Dr. John C. Lilly, a highly regarded neurophysiologist at the National Institutes of Health, who conceived a series of extreme experiments to test ideas about gaiting our sensations: in other words, stopping them from reaching consciousness.

Lilly's experiments built on Giuseppe Moruzzi and Horace W. Magoun's discovery in 1949 that our brainstem regulates our consciousness. Scientists like Lilly then wondered if our sensory organs bombard us with sensations to keep us aroused and conscious. Lilly wanted to know what would happen if we stopped this bombardment. He conceived of a way to allow only the smallest trickle of sensations to reach the thalamus, inventing the isolation tank, in which you are suspended in buoyant body-temperature water in total darkness and silence. The tank induces the feeling of floating in space, free of all bodily sensations.

Lilly himself was the first, and for years his principal, study subject in this experiment. He spent hours in his "sensory deprivation" chamber. He found he did not fall into sleep but was transported to "waking dreams" and "mystical states," although he always knew who and where he was.

During this period, Lilly suffered a near-fatal accident that had nothing to do with his work. He was giving himself an intravenous antibiotic injection for a minor illness when something went terribly wrong: air bubbles had entered the syringe and, paradoxically, somehow cut off the blood supply. He collapsed in a coma, which lasted for hours.

It is clear from his subsequent account that during Lilly's coma he had a near-death experience. He entered an unearthly realm of golden light that stretched to infinity. He felt peace and bliss; two guardians telepathically expressed an overwhelming sense of love and assured him that when he returned to his body he would someday be able to perceive the oneness of all things.

After he recovered, Lilly was determined to return to that blissful place. He knew that sensory deprivation had carried him toward but short of that goal, and he wanted to unlock the door to mystical union. He decided to use his isolation tank in conjunction with a newly discovered chemical.

In 1943, chemist Albert Hoffman accidentally ingested an intoxicating amount of a compound he had just synthesized and invented—lysergic acid diethylamide, or LSD. Hofmann continued the legacy of scientists self-experimenting in the spiritual realm and took a large dose of the drug the next day. He recorded in his journal many of LSD's effects, which are commonly known today. After the war, Sandoz, the manufacturer of LSD, saw the drug's potential and made it widely available to researchers like Lilly.

With an unlimited supply of pure LSD on hand, Lilly opened a mystical door by injecting himself with high doses of the substance and then entering his isolation tank. Now he could travel to other spaces without the fear of death. In the beginning of his journeys, Lilly sought and found the guardians from his NDE and learned that they were primal forms of human consciousness. Fear was a prominent part of his early explorations—not fear of death but fear of a "psychosis" from which he would not be able to return. He left his body and became a "single point of consciousness" where "one's self still exists," a state of mind that has strong similarities to the experiences of Stace's introvertive mystics. Lilly encountered an ineffable vastness and was transported to universes of immense energy and fantastic light inhabited by conscious beings. After many trips he was forced to return his stockpile of LSD to Sandoz because the federal government criminalized the drug, which effectively ended human hallucinogenic research.

Lilly left the NIH and continued his isolation tank experiments using ketamine instead of LSD. Then a new idea captured his imagination—to communicate with dolphins, social creatures with brains larger than our own, who were capable of complex communication through sound, gave themselves names, and recognized themselves in a mirror.

James and Mescal

Lilly was absolutely convinced that what he experienced in his isolation tank under the influence of LSD was real. "I knew that this was the truth," he wrote years later.

Were Lilly's experiences authentically spiritual?

Absolutely, if we apply James's criteria. "By their fruits ye shall know them, not by their roots," he wrote. In other words, one should judge the value of any spiritual experience by its lasting effects, not its cause.

James himself briefly experimented with hallucinogens. He took mescal at the urging of America's most eminent neurologist of his period, S. Weir Mitchell. He wrote about the experience to his brother Henry:

> I had two days spoiled by a psychological Experiment with *mescal*, an intoxicant used by some of our S. Western Indians in there [sic] religious ceremonies, a sort of cactus bud, of which the U.S. government had distributed a supply to certain medical men including Weir Mitchell who sent me some to try. He had himself been "in fairyland." It gives the most glorious visions of colour—every object thought of appears in a jewelled splendor unknown to the natural world. It disturbs the stomach somewhat but that, according to W.M., was a cheap price, etc. I took one bud 3 days ago, vomited and sputtered for 24 hours and had no other symptom whatever except that and the Katzenjammer the following day. I will take the visions on trust.

James had a slightly different take when he wrote to his friend Benjamin Paul Blood about his mescal experience. Blood had vigorously advocated using nitrous oxide ("laughing") gas, and his encouragement had led James to experiment with the anesthetic. James was unabashed about the insights he gained under the gas's influence: "I strongly urge others to repeat the experiment."

James wrote to Blood after ingesting mescal: "I have just been having an amusing experiment in seeking truth by intoxication. Not a flicker of light or colour, not a twinge of rationality, only, loathsome sickness the whole time. Should you like me to send you some? It might affect *you* less strongly in that way!"

James did not turn his back on the ancient and highly developed practices in indigenous cultures of using hallucinogens such as mushrooms and peyote for spiritual purposes—he openly embraced them. That same strategy is bearing scientific fruit today.

The New Experiences

Before the crackdown on using LSD for research, Walter Pahnke, a Harvard Divinity School graduate student, conducted the "Good Friday Experiment," showing that psilocybin could have a lasting positive impact on the spiritual lives of healthy people. Theological seminary student volunteers received psilocybin during a Good Friday religious service: the spiritual effects persisted for as long as twenty-five years.

Although Pahnke's experiment elevated mystical drug research above the anecdotes of Lilly and others, it remained a far cry from using spiritual drugs as molecular probes into the basis of consciousness. In 2006, Dr. Roland Griffiths and colleagues at Johns Hopkins showed the durability of drug-induced spiritual experience in thirty-six mostly middle-aged, well-educated, well-adjusted subjects who had never experimented with hallucinogens. They had varying spiritual orientations, but each—at least intermittently—attended religious services and engaged in prayer, meditation, or spiritual discussions.

Griffiths gave his subjects psilocybin in a warm and comfortable living room–like setting. They were encouraged to lie down on the couch, use eye masks, and listen to classical music with headphones to block out distractions (these sensory isolation methods were considerably less extreme than Lilly's tank). Most subjects had a "complete" mystical experience according to the "M" scale, a sense of unity indistinguishable from the mystical experiences of Plotinus, Eckhart, and Saint Teresa. The drug's impact reverberated long after it had been cleared from the brain. Two months after the experience, two-thirds of the subjects ranked it as one of their life's top five most meaningful events, on a par with the birth of a child or the death of a parent. Eighteen months later they still ranked it as one of their life's most important experiences, reporting it had improved their sense of well-being and positively changed their lives.

A Molecular Scalpel

Experiments like Lilly's and Griffiths's give us a molecular scalpel that identifies the physical brain structures and physiological brain processes essential to spiritual experiences. LSD and psilocybin strongly mimic the actions of the naturally occurring brain chemical serotonin. Serotonin plays an important role in a huge array of brain functions and disorders, including stress, depression, anxiety, memory, and attention. Serotonin also shapes our brain's responses to novel and threatening situations.

Serotonin is the last of the three neurochemical systems that we are examining, along with acetylcholine and nor-Adrenaline, that regulate consciousness. Serotonin nerves cluster near the locus coeruleus in the brainstem—in an area called the dorsal raphe, which is slightly larger than the locus coeruleus and is also important to consciousness. Serotonin nerves project through the major brain regions, including the limbic system. The serotonergic nerves act in tandem with nor-Adrenaline, balancing waking consciousness against the REM sleep of the cholinergic nerves.

Many clues point to serotonin's role in spiritual experience. For example, antidepressant drugs like Prozac, which works through the serotonin system, interfere with nearly all of LSD's effects, including the psychic ones.

Once released by the nerve, serotonin affects only other nerves that have a specialized molecular receptor. The released serotonin binds to the receptor, unleashing a cascade of chemical reactions that change how the nerves act and react to one another.

There are many types of serotonin receptors. The most important for mystical experiences appears to be the type known as serotonin-2. Both LSD and psilocybin molecules bind tightly to serotonin-2 receptors and cause nearly identical neurochemical actions and mystical experiences.

It's telling that a research drug that blocks serotonin-2 receptors, ketanserin, also blocks psilocybin's mystical effect. Administering this drug to someone is like physically cutting out the part of the brain that causes mystical experiences.

Our understanding of serotonin-2 brain chemistry is still in its inception. These receptors are being investigated for their role in depression, schizophrenia, stress, pain, and fear. Serotonin and its receptors are not just in the brain. In fact, most serotonin is found outside of the brain and helps regulate organs like the heart, lungs, and gut. Serotonin-2 receptors are even found in the hand's blood vessels, which may be why our hands feel cold and clammy when we're afraid.

Following the trail of serotonin-2 receptors promises to lead us to how the brain works in mystical experience. Inside the brain, serotonin-2 receptors are unevenly distributed, and their whereabouts are intriguing. They are heavily concentrated in the limbic system, including the hippocampus and the amygdala. But exactly where serotonin-2 receptors exert their mystical effects remains to be seen.

One of the tried-and-true ways of localizing a brain function is to remove a part of the brain and see if the function is lost. In the early 1960s, such an investigation was undertaken to discover brain regions important to mystical experience (today's ethical standards would not permit this kind of research). For his Ph.D. thesis, Dr. Eustace Serafetinides in London gave LSD to twenty-three epileptic patients who were having part of their temporal lobes surgically resected (including the amygdala and the hippocampus). After the surgery, Serafetinides gave the subjects a second dose of LSD. He found that the effects of the drug were considerably blunted, regardless of whether it had been the right or the left limbic structure that had been removed. This is strong evidence that somewhere in the amygdala or the hippocampus lies circuitry important to mystical experience.

But the amygdala and the hippocampus are not likely to be the whole story. More recently, for example, PET scans in subjects given psilocybin have shown that along with the temporal lobe, psilocybin strongly activates the limbic system's anterior cingulate region. Scientists are now poised to use the more precise MRI scan to find where psilocybin makes its mark. It would not surprise me if the anterior cingulate turns out to be important to the mystical effects of psilocybin.

Another exciting research technique offers great promise in locating the mystical in the brain. Serotonin receptors can be radioactively tagged, and their location revealed by PET brain scans. In an early study, brain scans showed higher limbic and brainstem activity of serotonin-1 receptors in people who have a stronger than average trait of self-transcendence and who accept an unseen world, believing, for example, in paranormal events such as extrasensory perception. It will be fascinating to see what is uncovered by tagging the serotonin-2 receptors engaged by psilocybin.

The questions of where and in what numbers serotonin-2 receptors influence mystical experience have yet to be fully explored, although the receptors have been tagged and human brains imaged in other conditions that could be linked to mystical experiences in ways that now seem bewildering.

The more someone weighs, for example, the fewer serotonin-2 receptors he or she has in the brain. In violent and aggressive people, serotonin-2 is greatly reduced in the prefrontal brain. Pain activates serotonin-2 receptors in limbic regions, including the orbitofrontal region we saw earlier, which assigns emotional value to pleasure and rewards, and the dorso-lateral, the problem-solving and planning part of the brain that is turned off in REM sleep. What all this could mean for mystical experience remains to be seen, but it probably means something.

The brains of men and women differ in many substantial ways. One of these differences turns out to be serotonin-2 chemistry. Men have more serotonin-2 receptors, especially in their left brain. In theory this provides the opportunity for more serotonin-2 action. But we need to know much more about the serotonin-2 puzzle before we can say more about brain chemistry, gender, and mystical experience. Sheer numbers of receptors alone may not fully explain the mystical issue. I have no empirical reason to suspect that between men and women serotonin-2 is different in the limbic regions that give rise to mystical experience. The larger number of serotonin-2 receptors in men may be more closely tied to other personality traits. From another vantage point, up to 50 percent of the difference between people in how strongly serotonin binds to serotonin-2 receptors is determined by genetic heritage. Someday this fact might help explain why some people, like Saint Teresa, Meister Eckhart, and Plotinus, have mystical experiences whereas others do not.

The role of serotonin-2 receptors in fear is fascinating. A sizable number of subjects in Griffiths's study had "strong or extreme" fear provoked by psilocybin, but the experience's positive personal meaning overshadowed this facet of it. Researchers pointed out that elaborate steps must be taken to prevent erratic or dangerous behavior from the fear that can erupt during a psilocybin experience.

The fear during hallucinogen use takes several forms: hallucinations, paranoid delusions, and panic attacks. I don't think this horror is simply a reaction to whatever bizarre thing seems to be happening to the user—the fear runs much deeper. The terror that can accompany LSD or psilocybin means that serotonin-2 receptors likely activate fear circuits directly, intertwining fear and mystical experience. Serotonin-2 receptors are integral to the heart's response to stress, especially when the threat is inescapable. Animals lacking serotonin-2 receptors also lack normal fear. Recently it was found that a chemical released by the brain during fear and stress enhances fear behavior in laboratory animals by sensitizing their brain's serotonin-2 receptors.

Other things connect fear and the serotonin-2 receptor. Low serotonin-2 activity in the limbic systems follows brain damage that renders someone incapable of feeling fear. Different genetic types of serotonin-2 make some of us vulnerable to panic attacks. People with higher serotonin-2 activity are more likely to have fearful personalities and avoid things that could be dangerous.

Neuroscientist Patrick Fisher and his group at the University of Pittsburgh undertook a fascinating study. They administered PET scans to healthy subjects and showed them pictures of angry and fearful faces. The scans revealed that serotonin-2 activity helps the medial prefrontal cortex regulate the amygdala, which is deluged with information during fear and when we experience the adrenaline surge of fight-or-flight. The medial prefrontal region, part of the limbic reward system damaged when the rod passed through Phineas Gage's brain, governs our visceral responses, shaping our reactions to fear; it may do so with the help of serotonin-2.

All this shows us that serotonin-2 receptors are an important component of the limbic circuitry and our experience of fear. It gives us a glimpse of how our survival instinct, fear, and the mystical are inextricably bound. These links may also be a way to explore the relationship between the mystical and the near-death experience.

Serotonin-2 receptors are going to play an increasingly important role in our future understanding of the mystical. They give us an important and effective tool for bringing science to bear on spiritual experience.

Spiritual Madness

There are obviously times when what passes for spiritual experience and its fruits is the product of a disordered or diseased mind. The mass suicide of Jim Jones and Jonestown springs to mind. But what about the Buddhist monks who immolated themselves to protest the Vietnam War? Was theirs a diseased spirituality?

One could argue that fundamentalism of any stripe is based on an illusionary and therefore disordered rigidity of belief, although the people who make those arguments can be as rigid as the fundamentalists they decry. In my discussion of consciousness and the self, I have tried to show that our supposed rock-solid sense of reality may be—from a neurologist's point of view, at least—as illusory as the spiritual realm is to the most devote atheist.

Spiritual experience is, by its very nature, beyond the border of our everyday lives, at least for most of us. How far beyond the border can the spiritual go before it becomes a disorder? This question can come up even when someone has what most people think of as an uplifting spiritual experience. This was at issue with Nancy, who wrote me:

> I would like your thoughts on a person who said that they died and saw God and was sent back (near-death experience). Are they to be believed or do they have a disorder? Please respond in detail as your answer is very important to me.

Are there instances when near-death experience should be treated as a disorder, a descent into madness? What criteria can we use to determine whether a spiritual experience of any type is healthy? Who's to say? The line between diseased and healthy spirituality can be imperceptible—not only to the diseased but to everyone else as well.

In the past, the subject of diseased spirituality and the brain was where neurologists focused their attention. After all, these were the types of cases most likely to come within their professional sphere as physicians.

Beyond the Fine Line

Frank, a friend of mine in college who eventually became a successful psychologist, experimented extensively with drugs in his twenties and then became an ardent spiritual seeker. I interviewed him to mine his spiritual experiences for this book, and he impressed on me the fine and permeable line between a healthy and disordered spirituality.

The first experience he described seems straightforward even if it is unconventional:

> In the wee morning hours, I was exhausted after a long night spent eating psilocybin mushrooms. Despite my weariness, I could still feel the mushrooms' stimulating effects as I lay still in bed with my eyes closed and began drifting off to sleep. Without any premonition, I became a point of consciousness completely devoid of any physical form. With the irresistible force of a black hole in space, I was rapidly pulled through successive layers of brighter and brighter light, leading to the center of being, the vortex of creation. I had a great fear that I would be lost if I reached the heart of this unbearably bright light. I was startled fully awake, and sat up with my heart pounding. It occurred to me then, and I believe it firmly now, that the layers of light insulated me day to day from the unbearable intensity that comes directly from an imminence, a pure consciousness.

This sounds like Saint Teresa's jeweled castle. By Stace's reckoning, Frank had a typical introvertive mystical experience, which occurred under the influence of psilocybin's serotonin-2 stimulation, just as his consciousness was shifting to sleep. Meister Eckhart would have recognized Frank's black hole of light to be the divine "spark" of God, the inner light all mystics see that is concealed within the shells of selfhood.

Frank said the experience was a precursor for an experience that happened a year or so later, which was so powerful and confusing that it nearly destroyed him.

> The incident I'm about to describe came when I was in grad school, studying philosophy. My friends and I were hanging out one night, smoking grass, and the stereo was playing. For some reason, someone put on Beethoven's 9th Symphony. As the music built, I was swept up by its power. This time, however, I did not merge into some kind of pure consciousness. The music was endowed with a divine, heavenly power. Its trumpets announced angels who had been sent to deliver a personal message from God to me. They telepathically told me I was on a divine mission: my fate was to be God's special envoy, bringing His truth of love and peace to earth. I had an absolute divine certainty that God had touched and chosen me to be his envoy.
>
> This feeling of being touched by God continued for the next couple of weeks. Each day I struggled with the intense feelings that I had touched ultimate reality, while my rational side was desperately trying to make sense of the experience. Gradually, bit by bit, I realized I was in the grips of a powerful delusion. My return to a cold reality without the warm comfort of divine certainty plunged me into unimaginable darkness and brought me the greatest challenge of my life: what could I ever believe after tasting divine certainty? How could I go on? I didn't just fall into the abyss—I *became* the abyss. I was terror-stricken that I would go mad, I feared for my life. I became withdrawn and apathetic and fell into a deep depression.

I asked Frank how he had recovered from this crisis. His answer stunned me.

> I realized that the only way I could survive was to find something I could believe in and there was nothing in conventional reality to guide me back home where I wanted desperately to be. All of my past beliefs and those of everyone else were shattered. So, I turned to the one truth that remained in my decimated world. I turned to that night of psilocybin mushrooms. I realized that the point of consciousness I had experienced was an absolute and immutable *certainty*. I had no choice; I had to survive. I started believing in myself, slowly at first, concentrating on the here and now. Soon, I was able to renew my focus on school. I switched from philosophy to psychology.

As Frank finished his story, I felt awe at the majestic power of mystical experience. It had brought him back from the brink. It was the bedrock on which he built his life after his crisis of faith. No doubt Meister Eckhart would have understood and approved of Frank's recovery as he reconstructed successive layers of himself around the mystical core of existence he had touched.

Joe's battle with the devil sparked my curiosity about the neurological underpinnings of spiritual experiences, but it was the gravity of Frank's descent into madness and the strength and clarity it took for him to pull himself back from the abyss that instilled in me a profound respect and compassion for the many types of spiritual experiences I have heard over the years—even if I'm not able to fully grasp their meaning.

Seizures and Spiritual Experience

For countless millennia, spirits and other apparitions have visited persons stricken by spells of bizarre consciousness and behavior, which have included staring blankly ahead, falling down, a failing of the limbs, and foaming at the mouth. These spells were interpreted as both demonic and divine. A well-chronicled story from the New Testament of Mark 9:14–29 describes Jesus commanding a demonic spirit to leave a "possessed" child who obviously suffered from seizures.

It's clear that even in the first century seizures had long been endowed with spiritual content, although not by everyone. Five hundred years earlier, Hippocrates had proposed a remarkably modern concept: "It is thus with regard to the disease called Sacred (seizures): it appears to me to be nowise more divine nor more sacred than other diseases, but has a natural cause like other afflictions." To Hippocrates this cause lay within the brain.

Nonetheless, neurologists suspect that there have been many people through the ages who had seizures as part of their spirituality, including Saint Paul, Joan of Arc, Saint Teresa, and Emanuel Swedenborg.

There are many seizure types, and the one that has been most readily tied to spiritual expression stems from abnormal electrical activity in the limbic structures of the temporal and frontal lobes. Limbic seizures can lead to brief but intense spiritual experiences. Some epileptics have spiritual delusions and hallucinations that can last days after the flurry of their seizures recedes.

Fear is the most common single isolated psychic manifestation of limbic seizures, but these manifestations can also include déjà vu, depersonalization (dreamlike experiences), out-of-body experiences, ecstasy, memory recall, hallucinations, and sensed presence.

Some epileptics with limbic seizures will become fixated on spiritual topics, usually outside of organized religion, although their epileptic spells seem to have no direct spiritual content. In some cases limbic seizures have led to a lasting spiritual conversion.

Limbic "ecstatic" seizures are the most distinctively spiritual of seizure types. Neurologists have long been fascinated by the ecstatic seizures of Fyodor Dostoevsky, which he describes in his fiction and autobiographical writings:

> I have really touched God. He came into me myself, yes, God exists, I cried, and I don't remember anything else. You all, healthy people he said, can't imagine the happiness which we epileptics feel during the second before our attack. I don't know if this felicity lasts for seconds, hours, or months, but believe me, for all the joys that life may bring, I would not exchange this one . . . might well be worth the whole of life.

Dostoevsky was not immune to his own suspicion that the spiritual ecstasy he felt was ignited by disease. "What if it is a disease? What does it matter that it is an abnormal intensity, if the result, if the minute of sensation, remembered and analyzed afterwards in health, turns out to be the acme of harmony and beauty, and gives a feeling, unknown and undivined till then, of completeness, of proportion, of reconciliation, and of ecstatic devotional merging in the highest synthesis of life?"

Sensing timelessness and touching ultimate certainty are certainly spiritual qualities, but nowhere have I found Dostoevsky to speak of feeling the mystical oneness in the writings of Plotinus or Eckhart. Dostoevsky's seizures could induce blissful euphoria, which he felt God had bestowed upon him, but is that any different from the effects of morphine on the brain? Other ecstatic seizures may produce the experience of mystical oneness, and we might be able to discern the brain areas responsible for the mystical by studying them, but so far we're not sure where they come from in the brain or what causes this particular seizure type. This is because they are so rare—rare enough that even their existence in Dostoevsky has been called into question.

What is more clear is that Dostoevsky is an archetype of the "temporal lobe personality," which is characterized by a preoccupation with religious topics and a compulsion to write, especially about religious themes.

Many neuroscientists lack the psychological or philosophical sophistication to be able to evaluate the spiritual experiences of some of the seizure patients they treat. They tend to lump mystical experience in with other types of experience that have a vaguely spiritual character. Michael Trimble and Anthony Freeman, neuroscientists from London, are notable exceptions. They applied Hood's "M" scale to patients with temporal lobe epilepsy and found that the epileptics did experience a loss of self and felt absorbed into something greater than themselves. But we are still left wondering if they had a mystical experience like Frank's fall into a singular point of consciousness or oneness or his subsequent delusion of being chosen as God's envoy. We still don't know if they were psychotic or mystical—and we don't understand, as far as the brain goes, how someone can be both.

Cases collected from neurologists around the world indicate that seizures with spiritual content tend to originate from right-sided limbic structures. In contrast, the spiritual psychosis and delusions after the electrical brain discharges have subsided more often occur when the limbic structures on both sides of the brain are diseased.

The rare historical figure aside, limbic seizures are not the catalyst for spiritual experiences in most people. They are important mostly because they can tell us what brain structures contribute to spiritual experiences, even if those structures are activated in most cases by other means.

Other Disorders of Divinity

The spiritual euphoria felt in a church, temple, or mosque may be neurologically indistinguishable from the euphoria of a narcotic (in contrast, experiencing the mystical oneness stands out because it is always seen as a spiritual experience). Religious and nonreligious themes often play through the same normal brain functions. A prophet speaking the words of God uses the same language areas in the left hemisphere that we use in our everyday speech. The "how" of his brain-formed language is well established by neuroscience, but "why" the prophet speaks lies beyond the brain, in the realm of faith.

In some instances, the grandiosity of believing you are Christ or a messenger of God may not be neurologically different from believing you are the incarnation of Caesar or Napoleon.

It is sometimes impossible to tell where the healthy spirituality of the sage ends and spiritual psychosis begins. It's clear when a vision of Christ compels a husband to kill his wife in order to save the world from Satan, rather than inspires him to establish a monastery. But between the extremes, a difference between healthy and diseased spiritual experience isn't always as clear.

Before Language

Exactly where in the limbic system mystical experience arises is a mystery, but I have confidence that we will know more about how it works in the brain in the near future. The mystical is not beyond language in a neurological sense. It is *before* language, residing in ancient brain structures concerned with our Darwinian survival. My strong hunch is that mystical experiences existed long before language came to our species. This is a rather startling thought. It means other animals aside from human beings may have mystical feelings.

Spiritual seekers of all kinds in all eras manipulate and fine-tune their instinctive arousal and limbic systems. Fasting (and the adrenaline surge it can bring on), meditation, living at high altitudes with low oxygen, and fear all bear on arousal and are tools used to step through the spiritual doorway in the brain, to enter the borderlands of consciousness, to fracture and fragment the self in the quest for transcendence.

It can, as said, be difficult to distinguish healthy from diseased spirituality. We are all too aware today of the power of spiritual experience to do harm, as we watch people blow themselves—and others—up in the name of God. We have seen that powerful feelings of fear and belief arise in the limbic system and our ancient and primitive brainstem, a part of the brain we share with other forms of life. The spiritual can be urgent and propulsive—spiritual truth has often come to be a matter of life or death. When we feel we connect with something larger than ourselves, we are willing to sacrifice our small selves to that something that we feel, that we know, is greater. This impulse can drive us to reach out to our fellow man in ways that are noble, inspiring, and even heroic. But it can also lead to intolerance and folly. In short, spiritual experiences bring out the best and worst in us.

At the same time, our pursuit of understanding them, of describing and discussing them, along with the rest of this enigmatic universe, is now simply a part of what it means to be human.

Epilogue

A NEW BIRTH OF WISDOM

"God is light, and in him is no darkness at all."

—1 John 1:5

Supposing spiritual truth were within dark energy and mass—what then? Since scientists, especially neuroscientists, know next to nothing about this energy and matter making up the vast majority of the universe, could we really expect neuroscientists to tell us much about the spiritual?

There are many good reasons to believe that the new neuroscience I have explored in these pages has left us in a cold, dark spiritual void. It may seem to many readers that we have reduced our peak experiences and most transcendent feelings and ideas to mere biology, and vegetative biology at that. Does this make us nothing more than accidental wisps of stardust between two eternities?

We have placed fragmented consciousness at the heart of many of our spiritual experiences and stripped away the illusion of the seamlessly integrated self. Odd as it may seem, we have shown that primal brainstem reactions seem to be at the root of the experiences that we think of as spiritual and that make us most human. This concept of "knee-jerk spirituality" deals a strong blow to the idea that free will is necessary to connect with whatever we feel is sacred.

At the neurologist's command, a flicker of electrical current to the brain makes it seem that our consciousness has been lifted from our body and is floating freely in space. The brain pathways used during "natural" spiritual experiences are the same pathways used by spiritual drugs, indistinguishable from otherwise genuine religious conversions, transforming lives long after the drug is flushed clear from the body. Clinical neurology tells us these are the same pathways distorted by some diseases of the brain that produce disorders fitting criteria for religious experience. Are spontaneous and authentic spiritual experiences nothing more than "experiments of nature" telling us how the brain works?

We have strong indications that much of our spirituality arises from arousal, limbic, and reward systems that evolved long before structures made the brain capable of language and reasoning. Neurologically, mystical feelings may not be so much beyond language as *before* language.

Given that we share many of the structures and systems in our brains with other creatures, we may not be the only primate with spiritual feelings. Great apes mourned their dead, and evidence suggests that Neanderthals believed in an afterlife. In fact, I strongly suspect that mystical feelings could exist in the many other mammals that are endowed with a limbic system that is very much like our own. And why can't dogs have out-of-body experiences?

These are challenging ideas. Our spiritual nature may not be the last bastion of human specialness, a particularly uncomfortable thought as advances in cosmology are magnifying our smallness and seeming insignificance in the face of a massive and mechanical universe.

Do these cold, hard clinical facts suck the divine nectar from our spiritual lives? My answer is an empathic NO! We are poised on the threshold of a new era that holds tremendous promise for a new level of spiritual exploration.

I urge caution, however. We need to be wary of our left hemisphere, the explainer and confabulator in our brains. It has led us astray so many times in the past, giving us a plethora of gods, including a mathematical god, to explain the natural world around us. The left hemisphere has given others the reasons to explain away our spiritual essence, often with hubris. It's unlikely that the left hemisphere today has changed its basic nature after more than one hundred thousand years of evolution. It is all too willing to look for natural and supernatural explanations.

I have faith that in my waking consciousness, when my dorsolateral brain is on, the laws of nature are immutable (at least for now). Apples do not fall upward from the tree; one plus one is never three. And I have faith in molecules. I witness ideas based on molecules saving lives every day I am in the hospital. Do I believe in molecules that I can't directly experience, things my left hemisphere tells me exist? Yes, I believe, until a better idea comes along. To me faith and science are not as different as some people have made them.

Basing one's spirituality on science is as foolhardy as basing one's science on spirituality. No matter if we could know *how* every single brain molecule makes spiritual experience, *why* the brain is spiritual will remain for many of us our most treasured mystery. There is room in the brain for faith. Separating the how from the why also creates an apparent paradox; the brain that brings us spiritual experience itself can be seen as spiritually neutral.

We have one certain fact at hand: we can be conscious. This is a great miracle no matter how it happens. I think I might get the philosopher David Hume to agree with me here, even though he considered miracles a "transgression of a law of nature," and this hardly seems to be a case for consciousness. So far we can say with near certainty that believing in experiences outside the brain is faith. Sensing that something is "more than coincidence" is also an expression of faith. It's folly to expect that science can prove or disprove the truthfulness of these experiences. But spiritual hope based on *false* science is cruel. The nature of faith makes it immune to science's demands for consensus, verification, and prediction.

Under the guise of science, researchers have claimed that near-death and out-of-body experiences "prove" that mind exists separate from the physical brain. Such a claim is the most extraordinary in all of science, surpassing even the dramatic assertion that other intelligent life exists in the Milky Way, our galaxy.

The Drake equation gives us the probability of contacting intelligent life in the Milky Way. It is a calculation based on the number of stars like our sun that are formed each year, the fraction of stars that have planets, the number of habitable planets, etc. The Drake equation rates the chances of other intelligent life out there as very small, but the point is that it is not zero. There is no such calculation for experience outside the brain—yet.

What does it mean to have a spiritual brain? What if there is a particular brain locale or system, a genesis for divine experiences like the mystical? And when we find it, will we try to nurture, destroy, or control it? I can imagine that the power of knowing how to stimulate the spiritual brain will bring temptations of the darkest sort. It is one thing to know the brain's machinery for language or how it puts sensation together, and quite another to be able to open it to searing experiences of ultimate truth or purpose. This knowledge might bring with it a Faustian deal with the devil, potentially transforming us from god to God. We would have the potential to provoke events such as Jonestown many times over.

What if we had a drug that acted on a specific part of the brain and caused us to experience the miraculous? Or imagine another drug, more refined than psilocybin, which precisely stimulated the mystical, bringing us to a state of "oneness," or even closer to God than we can now imagine.

As doctors, how should we use that drug? In what crisis should divinity be administered? If a patient has terminal cancer, physicians are pledged do all that is possible to alleviate suffering: to withhold narcotics such as morphine from the dying when this relief is desperately needed and sought is a breach of professional ethics. Should the same obligations apply to drugs that stimulate the divine? Would withholding oneness or enlightenment be inhumane if we had the power to summon it at will?

On the practical side, maybe we should routinely offer rapture, like we offer resuscitation for cardiac or respiratory arrest. I'm not thinking of rapture of a particular creed but intense spiritual bliss. Health care workers are required to ask patients or their families if they desire resuscitation should their heart or breathing cease, and placing "Do Not Resuscitate" labels is routine on hospital charts. Add to these labels "Rapture" or "Do Not Rapture." "Bring on the divine rapture" you tell your doctor, or "No rapture for me, thank you."

Morphine clouds thinking and alters judgment. It mimics natural brain molecules to make one feel unnaturally good. It's part of the comfort we give to alleviate the pain and anxiety that often accompanies death. In fact, if a newly designed narcotic relieved terminal cancer pain without the accompanying euphoria, we would have a dilemma. Should we withhold morphine and its euphoria and thereby fail to fully relieve the suffering of a dying psyche? Whether to induce the experience of the divine is a decision too important for medicine to make alone.

I can see these possibilities and so much more, glimmering ahead, still out of reach but getting closer. We are all of this world, and my experience optimistically compels me to believe that understanding the brain as a spiritual organ strengthens our quest for meaning and complements a mature spirituality. My deepest hope is that this quest will ultimately bring us to a new birth of wisdom.

NOTES

Page

24. *Special MRI scans.* Blood oxygen level dependent (BOLD) or functional MRI (MRI).
24. *It takes more brain activity to work out if a statement is false then it does to decide it's true.* The authors point out that Spinoza's conjecture might be neurologically true, that understanding a statement is tacit acceptance of its truth.
24. *nineteenth-century intellectual and theologian.* Henry James Sr. was a staunch advocate of the theologian Emanuel Swedenborg of Sweden who, in the 1740s, was among the first persons to localize function in the brain. He attributed human intellectual faculties of imagination, memory, and thought to the frontal lobes. Swedenborg's pioneering ideas remained obscure partly because he ended his scientific theorizing and turned his attention to religious matters after several spiritual experiences.
26. *popular nineteenth-century "spiritualist" movement of mediums and clairvoyants, which was rife with charlatans.* Spurring the magician Harry Houdini to wage a vigorous campaign against their practices.
27. *drawing heavily on spiritual experience that had been documented by psychologists of religion who were his contemporaries.* This group includes his former student Edwin Starbuck and their works are worth reading for their own merits.

28. *John Addington Symonds was a renowned Victorian art historian and Renaissance scholar.* Most likely Symonds came to William's attention through his brother Henry, who wrote "The Author of Beltraffio," a short story based on the tragic marriage of his friend Symonds.
29. *Symonds's experiences is one of the first times a neurologist with a modern understanding of the brain grappled with the spiritual, and he does it badly.* James himself was perplexed by this vitriolic and very public assault by a respected physician under the guise of medicine. He wrote that Symonds's "critic gave no objective grounds whatever for his strange opinion."
32. *I examined her more closely for infantile reflexes that are suppressed by the frontal lobe when we mature.* Such as suppressing eye blinks when the forehead is lightly tapped repeatedly.
33. *"Although the soul sees this for a certain length of time, it can no more be gazing at it all the time than it could keep gazing at the sun."* "Image" of God through Jesus Christ.
44. *Sherrington, who would eventually win a Nobel . . .* 1932.
45. *It was Ramón y Cajal, Sherrington's inspiration, who found the convincing evidence that the nerve cells, the living units of the brain, were not physically connected but led independent lives.* Camillo Golgi's work staining nerve cells so they could be seen under the microscope was the starting point for Cajal's contributions, and both shared the Nobel Prize. Strangely, Golgi never believed in Cajal's principle of independent nerve cells and remained a bitter critic to the end.
53. "*milieu intérieur*" Claude Bernard.
59. *"The most common lie is that with which one lies to oneself."* The Delphic Oracle proclaimed that no one was wiser than Socrates. Nietzsche criticized Socrates as much as he admired him, yet in important ways they were very much alike. Both felt that a person could approach some form of truth. Socrates, who embraced the oracle's wisdom about self-knowledge, thought that truth rested on the "Good" or as Plotinus later said the "One." Nietzsche also believed in a truth, for he respected the liar as someone who knows an inner truth. This truth was based on Nietzsche's idea of "will to power," a will that he believed could overwhelm and transcend the self's will to survive.
59. *the self . . . hotly debated and now awesomely arcane.* As only Nietzsche could remark: "A matter that becomes clear ceases to concern us. What

was on the mind of that god who counseled: 'Know thyself!' Did he mean: 'Cease to concern yourself! Become objective!' And Socrates? And 'scientific men'?"

64. *Synapses in the left temporal lobes . . .* Hippocampus.

64. *A sense of agency and the "me" of experience can persist.* See Oliver Sacks, "The Lost Mariner."

66. *brain structures critical to her forming new memory.* Amygdala hippocampus and their connections.

70. *the specialized nerve circuitry . . .* Neurologists loosely use the words "circuit" and "wiring" when referring to specialized cortical regions and their connections, although they bear no physical resemblance to electronic circuits or wires.

70. *Other disconnections may occur . . .* Norman Geschwind at Harvard in 1965 sparked a reawakening of the ideas of disconnected brain syndromes that was first conceived in the late nineteenth century. Since Geschwind, the study of disconnected syndromes has again fallen out of and more recently returned to favor.

71. "*. . . or I'm nuts.*" When injury strikes the left parietal lobe that controls the right arm no such denial occurs as far as we can tell since language is lost at the same time.

75. *completely paralyzed except for the ability to move her eyes up and down.* Similar to Jean-Dominique Bauby's stroke chronicled in *The Diving Bell and the Butterfly* and the stroke of Monsieur Noirtier de Villefort in Dumas's *The Count of Monte Cristo.*

77. *We now know that this area is important for rational decision-making and processing emotion.* Injury to frontal lobes does more than damage how we experience our own mind: it also impairs the ability to infer the mental content of other people's minds, making the frontal region essential for empathy. The detection of a mind in other people is sometimes called the "theory of mind" since we can directly experience only our own mind, and must presume there is a mind in others.

77. *Both sides of the brain draw on nearly the same autobiographical memories.* Although the left memory region (hippocampus) is dominant and this is the side of the brain that talks, hippocampal regions on both sides communicate with each other.

81. *Briefly anesthetizing one side and then the other side of the brain.* Wada test.

84. *the connection between right and left emotional (limbic) structures was intact.* Connected by the anterior commissural fibers.

84. *Or could one side be wracked with guilt while the other self-righteously trumpeted virtue?* This is based on the assumption that the connection between the emotional parts of the brain, unlike in the clinical cases cited, could be severed.

87. *Dr. Jules Cotard . . .* Whether Dr. Cotard served as the literary model for Marcel Proust's Professor Cottard is disputed.

88. *the patient is convinced that he or she does not exist, in the face of clear contradictory evidence.* Impervious to the force of Descartes's dictum "*cognito ergo sum*," "I think therefore I am."

95. *as many as eighteen million Americans may have had one, according to a 1997 issue of* U.S. News & World Report. Eighteen million may be a conservative number. In the early 1980s, Gallup found that 15 percent of the people his team polled had in their lives been on the verge of death or had a close call and with it also had an "unusual experience." If we apply this figure to today's population of 300 million, then 45 million Americans may have had some kind of unusual experience associated with near-death.

96. *when his neck found its way into the jaws of his tiger, Montecore.* Injuring his carotid artery.

98. *I was given medication to speed it up and stabilize its rhythm.* Atropine for the slow heartbeat and lidocaine for heart rhythm.

104. *I would score Seymour's near-death experience eight.* Jane Seymour's Greyson score is an approximation since the person has to answer the questions for themselves to make the score valid. Symonds earned himself a fifteen in my estimation.

104. *She had had a brain aneurysm as a young woman.* Weakening of the artery that may burst and cause a stroke.

106. *Ayer . . . spent most of his professional career teaching at the University College London and Oxford.* Ayer later regretted that his original Shakespearian title "That Undiscovered Country" was not retained.

106. *He wrote on William James . . .* Ayer also delivered the Gifford Lectures, the same lecture series that earlier formed the basis of both William James's *The Varieties of Religious Experience* and Charles Sherrington's *Man on His Nature.*

106. *and produced a biography of Britain's foremost contemporary philosopher, Bertrand Russell.* Russell had famously spoken out against religion, choosing to side with logic over mysticism.
107. *Nearly all of our cells are replaced every seven years. And even the molecules are exchanged that make up the cells that remain with us throughout our lives.* Early in his career, Ayer found James's memory ideas fell short of explaining the continuity of self over time.
108. *"we should still have no good reason to regard ourselves as spiritual substances."* Whether Ayer admitted in private that he movingly encountered a "Divine Being" is disputable.
109. *for purposes of divination and to explore his subconscious.* Translations of Jung use the word "unconscious" rather than "subconscious" and should not be confused with the way a neurologist refers to an unconscious state or coma.
115. *Memory is the first brain function . . .* Along with the hippocampus in the temporal lobe.
116. *Extraordinary claims, however, require extraordinary evidence* When it comes to believing in the paranormal, I start with the yardstick provided by David Hume on miracles that violate the laws of nature: I believe in the paranormal only if *not* believing would mean having faith in something even more miraculous.
119. *In these studies, researchers found no underlying conditions that trigger near-death experiences . . .* Recent investigations of near-death experiences have examined each case consecutively, which is the preferred scientific method.
120. *a chemical used by nerves to communicate with one another, which excites the brain.* Linked to the drug phencyclidine (PCP).
129. *It turned out that a lot happened in those fifteen seconds.* Lambert, along with his colleague Earl Wood, made detailed observations of three hundred men who took a total of 9,500 trips in the centrifuge.
130. *When not enough blood is pumped to the head, the eye fails first, causing tunnel vision before the brain fails and unconsciousness occurs.* A model of Lambert's centrifuge remains at the Mayo Clinic (Scottsdale) today. I thank Dr. Vanda Lennon for her guidance.
134. *Descartes thought the pineal, like other glands, distributed "animal spirits" within the brain's cavities, swirling in currents and eddies that caused muscles to move the limbs, eyes, and face.* Like his many philosophical and

mathematical ideas, Descartes's thoughts about the pineal gland were strikingly original. He was certainly wrong that the pineal gland is the brain's centralized spiritual-material transducer, although, oddly, because of its central location, the pineal gland turns out to, in fact, be an interesting neurochemical transducer. Originally a light receptor ("third eye") in early vertebrates, the pineal evolved into the gland that rhythmically synthesizes melatonin, the hormonal messenger of darkness. The pineal gland is an essential cog in our brain's master clock, but the pineal gland is itself not the master clock. Instead that role goes to the suprachiasmatic nucleus situated just above where the nerves from the eyes cross in the brain. Descartes believed that keeping time is a fundamental brain attribute, which he naturally assigned to the pineal gland.

135. *Wilder Penfield . . . devoted his neurosurgical career to exploring the human cerebral cortex by stimulating it in hundreds of completely awake patients to see what happened.* From these experiments he was able to develop the homunculus brain map for movement and sensation that was important in our discussion of phantom limbs.

143. *a small trickle of electrical current . . .* The current used to cause out-of-body experiences is less than five-thousandths of an ampere or roughly one hundred times less than what a sixty-watt lightbulb draws.

143. *Not knowing the devil in the all-important details of how the study will be conducted limits what I can say at this point.* The investigation needs to be examined by scientific peers before it is discussed in the media.

147. *Here, too, a more worldly explanation is not only likely but obvious.* Over twenty-five years and in countless cases, I have monitored physiological signals including EEG and brainstem responses during surgery. It is quite possible to have brain activity while the standard EEG is "flat" in the electrically hostile operating room where the brain waves, millionths of a volt in power, are overwhelmed by stray electrical signals from the many electrical devices necessary for surgery.

150. *Surprisingly, the results showed the experience in both groups to be almost identical.* Unfortunately the investigators did not list the specific causes of the "fear-death" and near-death experiences. They could have confused the two experiences if they considered fainting harmless. From the brain's perspective, nearly identical brain reactions are triggered in

fainting and cardiac arrest; they differ mostly by how long the blood flow is reduced.

155. *the source of nearly all the brain's nor-Adrenaline . . .* Nor-Adrenaline is an acceptable substitution for nor-epinephrine, the more widely used term in neuroscience.

155. *These cells are darkly pigmented. They stand out to the naked eye, and they send microscopic projections to nearly every recess in the brain.* The basal ganglia are an exception.

158. *When our attention is calmly focused, thousands of locus coeruleus neurons fire in synchronized bursts . . .* Synapses are bypassed and the nerves use a special connection called a gap junction.

164. *Cannon spent much of his career arguing against the "James-Lange" theory of how the body expresses emotions, especially fear.* This theory had been the basis of Cannon's thesis when he was James's student.

164. *Cannon discovered that the adrenal glands secrete a hormone, which he called adrenaline.* Also known as epinephrine.

165. *this adrenaline surge.* In posttraumatic stress disorder the adrenaline surge runs amok.

165. *". . . serve to affect muscular power and mobilize sugar for muscular use—thus in wild state readies for fight or run!"* Immediately Cannon's critics called the reaction to danger "fight or flight," which harkened back to an earlier work where William James used the phrase to describe the response to seeing a bear.

171. *what this experiment suggests is that we* perceive *time slowing or our thoughts speeding up when we're in danger.* Surviving falls from mountains, hot air balloons, and other heights is chronicled by a Web site called the Free Fall Research Page (www.greenharbor.com/fffolder/ffresearch.html).

174. *The tight interconnection between the amygdala and the hippocampus (and other brain structures) . . .* Cholinergic basal forebrain nuclei.

179. *the amygdala enables the medial prefrontal brain to make sense of our visceral responses, setting the moods that guide our actions and choices* And in a depressed mood this area may be overactive. The reward system may prove to be a new avenue for treating depression.

179. *the "somatic marker" that gives our subconscious visceral warnings to avoid certain behaviors.* Maybe James was not so far off and the body's response, like a racing heart on seeing a bear, is important to our full emotional experience.

180. *Just the sight of chocolate is enough to make the orbital prefrontal brain of chocolate cravers "light up" on an MRI.* And connecting regions like the anterior cingulate cortex.

188. *Wave after wave ripples through the visual brain for as long as we are in REM.* Our sleeping brain cycles between non-REM and REM stages every ninety minutes. Most sleep is non-REM, with REM making up only about 25 percent of sleep, usually in four to six discrete periods over the last one-third of the night.

188. *Our eyes and breathing muscles are left unaffected.* In REM behavior disorder, spinal paralysis fails to weaken the limbs during REM sleep, often leading to injury. This condition presages Parkinson's disease.

189. *Although the purpose of REM sleep and dreaming remains a nearly a complete enigma, we do know that REM sleep is a vital function.* REM sleep occurs only in mammals and birds.

189. *It has a role in delaying gratification.* Nietzsche thought of human nature as a struggle between two forces: Apollonian reason and order wrangling with Dionysian passions (that became restrained and harnessed passions in his later writings). I am tempted to view part of this struggle as one between the dorso-lateral Apollonian brain and the orbital prefrontal Dionysian brain.

197. *Patients with narcolepsy have a chemical deficiency . . .* Orexin from the hypothalamus.

197. *a portion of the REM switch called the vlPAG.* Ventrolateral part of the periaqueductal gray.

199. *When David Livingstone . . .* Of the quotation "Dr. Livingstone, I presume?"

200. *With the help of an Internet community committed to providing information and support to people who have had near-death experiences . . .* Near Death Experience Research Foundation (www.nderf.org).

202. *This difference in the arousal brain is likely close . . .* Physically or functionally.

207. *When the vagus nerve is stimulated for medical purposes in patients . . .* To treat seizures that aren't responding to medications.

207. *Near the REM switch . . .* Lateral parabrachial region.

213. *The acetylcholine nerves instrumental to REM consciousness connect to the brain's reward center.* The ventral tegmental reward region uses the neurotransmitter dopamine and sends fibers to the limbic system including the nucleus accumbens, prefrontal cortex, amygdala, and hippocampus.

215. *Unfortunately, no researcher has yet captured fainting and REM consciousness on physiological recordings . . .* Forster and Whinnery offer recordings showing REM, but these tracings are too rudimentary for us to draw even preliminary conclusions. It is a major challenge to record REM's physiological footprints during syncope through the shroud of high amplitude brainwaves caused by inadequate brain blood flow, combined with the inner ear's domination of eye movements.

218. *I don't see REM intrusion as the last word on near-death experience, nor REM consciousness as the last word on spirituality.* Is the REM system activated to dampen the locus coeruleus in danger? Can REM be triggered by the vlPAG when blood pressure is low? These are but a few of the many remaining unknowns.

222. *They had slightly more REM intrusion than our other subjects, but not enough to be statistically significant.* All of these preliminary suspicions must be viewed cautiously since they have not been subjected to peer review.

224. *we did ask our subjects if they felt united or harmonious with the universe* We asked: "Did you feel a sense of harmony or unity with the universe?" (from the Greyson scale).

224. *Although James's criteria help us understand mystical experience, they don't give us a means to scientifically measure it.* With sufficient sensitivity and specificity.

225. *that process can be studied with psychological methods.* Until very recently, the psychology of religion had lost vigor in part because, like neurology, psychology has also deemphasized the importance of subjective experience.

228. *Perceiving unity is the most important thread running through both the extrovertive and introvertive mystical experiences.* See table on page 299.

228. *His spiritual experiences are not conveyed in a familiar narrative style . . .* Ancient philosophic and other early writings on personal spiritual experience lack the first-person narrative so familiar today. Oftentimes they speak in metaphors that translate poorly. Bringing to life the direct mystical experience underlying Plotinus's philosophy, late nineteenth-century author R. A. Vaughan reproduces a letter from Plotinus to his aspiring disciple Flaccus on the nature of mystical experience. This letter is often cited in modern literature, including Stace and Bucke, but unfortunately it appears to be completely fictitious. Furthermore, Plotinus's most widespread quote, "flight of the alone to the Alone," is a stylized translation that risks being misleading (VII.9.11).

230. *strikingly similar to the Eastern mystics.* Sankara.

230. *The unity experience brought Eckhart into serious conflict with Church authorities.* Part of the basis for his charge of heresy was the timeless sensation of mystical experience, for touching the timeless upset the Church as well.

235. *during the experiment the nuns did not achieve any of the core features of a mystical experience, including the loss of time, space, and feelings of unity.* It is unclear which fifteen questions the nuns were asked out of the thirty-two questions on Hood's "M" scale. The nuns scored high on only three questions, including "something greater than my self seemed to absorb me," "profound joy," and "an experience which I knew to be sacred."

237. *to test ideas about gaiting our sensations: in other words, stopping them from reaching consciousness.* Lilly had developed a novel electrical waveform to stimulate tiny brain regions harmlessly for long periods that is used by researchers even today. Yet it is on brain stimulation of a different sort that he left his deepest mark.

238. *He felt peace and bliss; two guardians telepathically expressed an overwhelming sense of love and assured him that when he returned to his body he would someday be able to perceive the oneness of all things.* I estimate Lilly scored twenty-four out of thirty-two on Greyson's scale.

239. *Now he could travel to other spaces without the fear of death.* Lilly encouraged a number of scientists at NIH to personally explore LSD's psychic effects. Many, including my mentor, graciously declined his offer.

240. *the Katzenjammer . . .* Hangover.

240. *James was unabashed about the insights he gained under the gas's influence . . .* "[Nitrous oxide] made me understand better than ever before both the strength and the weakness of Hegel's philosophy."

243. *The most important for mystical experiences appears to be the type known as serotonin-2.* Specifically, serotonin type 2a (5-hydroxytryptophan-2a or 5-HT2a). Psilocybin also has serotonin type 2c actions, but these seem less important to its mystical properties.

243. *They are heavily concentrated in the limbic system, including the hippocampus and the amygdala.* Serotonin-2 receptors are also concentrated in the cholinergic cell clusters and other regions affecting REM sleep. This suggests some role in consciousness (such as arousal), yet the action of serotonin-2 on REM sleep is not clear.

243. *Dr. Eustace Serafetinides in London gave LSD to twenty-three epileptic patients who were having part of their temporal lobes surgically resected (including the amygdala and the hippocampus).* Usually there are fewer psychological effects than you might expect from this surgery, so long as the side that remains is healthy.
244. *regardless of whether it had been the right or the left limbic structure that had been removed.* It is very difficult to determine if the right, left, or both limbic structures are important to LSD's effects.
244. *psilocybin strongly activates the limbic system's anterior cingulate region.* Psilocybin also activates limbic regions in the frontal lobes, while at the same time inactivating the thalamus.
244. *In an early study, brain scans showed higher limbic and brainstem activity of serotonin-1 receptors.* Serotonin-1 a.
245. *how strongly serotonin binds to serotonin-2 receptors is determined by genetic heritage.* In yet another twist, psilocybin both activates and partially inactivates the serotonin-2 receptor.
251. *Nonetheless, neurologists suspect that there have been many people through the ages who had seizures as part of their spirituality, including Saint Paul, Joan of Arc, Saint Teresa, and Emanuel Swedenborg.* Jeffrey Saver and John Rabin give a superb review of spiritual experience and brain disease from a classic neuroscience stance.
252. *so far we're not sure where they come from in the brain or what causes this particular seizure type.* Although they have not been well localized in the limbic system, it would seem they activate the reward system.
257. *neuroscientists, know next to nothing about this energy and matter making up the vast majority of the universe, could we really expect neuroscientists to tell us much about the spiritual?* A nod to Friedrich Nietzsche.
261. *Add to these labels "Rapture" or "Do Not Rapture." "Bring on the divine rapture" you tell your doctor, or "No rapture for me, thank you.* In actual practice we would have to use another word, perhaps ecstasy or grace. The standard DNR abbreviation for "do not resuscitate" would be too easily misinterpreted as "do not rapture," certainly an error to avoid, because someone might not want to be resuscitated but would want the ecstasy of divine presence at or near death.

REFERENCES AND RESOURCES

PROLOGUE

Hobson, J. A. R.W McCarley, P. W. Wyzinski. Sleep cycle oscillation: reciprocal discharge by two brainstem neuronal groups." *Science* 189 (1975): 55–58.

Moody, R. *Life After Life.* Covington, GA: Mockingbird Books, 1975.

CHAPTER 1

Austin, J. H. *Zen and the Brain: Toward an Understanding of Meditation and Consciousness.* Cambridge, Mass.: MIT Press, 1998.

Bucke, R. M. *Cosmic Consciousness: A Study in the Evolution of the Human Mind.* Philadelphia: Innes & Sons, 1901.

Coe, G. A. *The Spiritual Life: Studies in the Science of Religion.* Chicago: F. H. Revell Co., 1900.

Crichton-Browne, J. "Dreamy Mental States." *Lancet* 3749 (1895): 1–5.

Crichton-Browne J. "Dreamy Mental States." *Lancet* 3750 (1895):73–75.

Hill, P. C., and R. W. Hood. *Measures of Religiosity.* Birmingham, Ala.: Religious Education Press, 1999.

Harris, S., J. T. Kaplan, A. Curiel, S. Y. Bookheimer, and M. Iaocoboni. "The Neural Correlates of Religious and Nonreligious Belief." Available at: http://www.plosone.org/article/info%3Adoi%2F10.1371%2Fjournal.pone.0007272.

Harris S., S. A. Sheth, and M. S. Cohen. "Functional Neuroimaging of Belief, Disbelief, and Uncertainty." *Annals of Neurology* 63 (2008): 141–147.

Immordino-Yang, M. H., A. McColl, H. Damasio, and A. Damasio. "Neural Correlates of Admiration and Compassion." *Proceedings of the National Academy of Sciences* 106 (2009): 8021–8026.

James, W., and M. E. Marty. *The Varieties of Religious Experience: A Study in Human Nature.* Harmondsworth, Middlesex, England; New York: Penguin Books, 1982.

Kandel, E. R. "A New Intellectual Framework for Psychiatry." *American Journal of Psychiatry* 155 (1998): 457–469.

Maslow, A. H. *Religions, Values, and Peak-Experiences.* New York: Viking, 1970.

Nair, D. G. "About Being BOLD." *Brain Research Review* 50 (2005): 229–243.

Saver, J. L., and J. Rabin. "The Neural Substrates of Religious Experience." *Journal of Neuropsychiatry and Clinical Neurosciences* 9 (1997) :498–510.

Stace, W. T. *Mysticism and Philosophy.* London: Macmillan, 1960.

Starbuck, E. D. *The Psychology of Religion: An Empirical Study of the Growth of Religious Consciousness.* London and Felling-on-Tyne: The Walter Scott Publishing Co., Ltd., 1899.

Symonds, J. A., and H. F. Brown. *John Addington Symonds: A Biography.* London: J. C. Nimmo, 1895.

Teresa, Peers E. A. *Interior Castle*, Image ed. New York: Doubleday, 2004.

Trimble, M. R. *The Soul in the Brain: The Cerebral Basis of Language, Art, and Belief.* Baltimore: Johns Hopkins University Press, 2007.

CHAPTER 2

Bernat, J. L., and D. A. Rottenberg. "Conscious Awareness in PVS And MCS: The Borderlands of Neurology." *Neurology* 68 (2007): 885–886.

Blanke, O., and J. E. Aspell. "Brain Technologies Raise Unprecedented Ethical Challenges." *Nature* 458 (2009): 703.

Burke, R. E. "Sir Charles Sherrington's *The Integrative Action of the Nervous System*: A Centenary Appreciation." *Brain* 130 (2007): 887–894.

Cline, D. B. "The Search for Dark Matter." *Scientific American* 288 (2003): 50–55, 58–59.

Crick, F. "Function of the Thalamic Reticular Complex: The Searchlight Hypothesis." *Proceedings of the National Academy of Sciences* 81 (1984): 4586–4590.

Finger, S. *Minds Behind the Brain: A History of the Pioneers and Their Discoveries*. Oxford and New York: Oxford University Press, 2000.

Fox, D. "Consciousness in a . . . Cockroach?" *Discover* 2007: 66–70.

Gaudin, S. "Nantech Could Make Humans Immortal by 2040, Futurist Says." Availableat:http://abcnews.go.com/Technology/AheadoftheCurve/immortality-nanotech-make-futurist/story?id=8726328.

James, W. *Psychology: The Briefer Course*. New York: Dover, 2001.

Jokl, E. "Sherrington, His Life and Thought." *Transactions and Studies of the College of Physicians of Philadelphia* 2 (1980): 223–235.

Merker, B. "Consciousness Without a Cerebral Cortex: A Challenge for Neuroscience and Medicine." *Behavioral and Brain Sciences* 30 (2007): 63–81; discussion 81–134.

Owen, A. M., M. R. Coleman, M. Boly, M. H. Davis, S. Laureys, and J. D. Pickard. "Detecting Awareness in the Vegetative State." Science 313 (2006): 1402.

Ribary, U. "Dynamics of Thalamo-Cortical Network Oscillations and Human Perception." *Progress in Brain Research* 150 (2005): 127–142.

Rumelhart, D. E., and J. L. McClelland, University of California San Diego PDP Research Group. *Parallel Distributed Processing: Explorations in the Microstructure of Cognition*. Cambridge, Mass.: MIT Press, 1986.

Sacks, O. W. *Musicophilia: Tales of Music and the Brain*. New York: Alfred A. Knopf, 2007

Saver, J. L., and J. Rabin. "The Neural Substrates of Religious Experience." Journal of Neuropsychiatry and Clinical Neurosciences" 9 (1997): 498–510.

Schiff, N. D., J. T. Giacino, K. Kalmar, J.D. Victor, K. Baker, M. Gerber, B. Fritz, et al. "Behavioural Improvements with Thalamic Stimulation After Severe Traumatic Brain Injury." *Nature* 448 (2007): 600–603.

Schiff, N. D., U. Ribary, D. R. Moreno, et al. "Residual Cerebral Activity and Behavioural Fragments Can Remain in the Persistently Vegetative Brain." *Brain* 125 (2002):1210–1234.

Schiff, N. D., D. Rodriguez-Moreno, A. Kamal, K. H. Kim, J. T. Giacino, F. Plum, J. Hirch, et al. "fMRI Reveals Large-Scale Network Activation in Minimally Conscious Patients." *Neurology* 64 (2005): 514–523.

Sherrington, C. S. *Man on His Nature*. New York and Cambridge: The Macmillan Company, The University Press, 1941.

Trimble, M. R. *The Soul in the Brain: The Cerebral Basis of Language, Art, and Belief*. Baltimore: Johns Hopkins University Press, 2007.

Zeman, A. "Consciousness." *Brain* 124 (2001): 1263–1289.
Zeman, A. "Persistent Vegetative State." *Lancet* 350 (1997): 795–799.
Zeman, A. "Sherrington's Philosophical Writings—A 'Zest For Life.'" *Brain* 130 (2007): 1984–1987.

CHAPTER 3

Assal, F., S. Schwartz, and P. Vuilleumier. "Moving With or Without Will: Functional Neural Correlates of Alien Hand Syndrome." *Annals of Neurology* 62 (2007): 301–306.
Bauby, J.-D. *The Diving Bell and the Butterfly*. New York: Knopf, 1997.
Berrios, G. E., and R. Luque. "Cotard's Syndrome: Analysis of 100 Cases." *Acta Psychiatrica Scandinavica* 91 (1995): 185–188.
Blackmore, S. J. *Consciousness: A Very Short Introduction*. Oxford: Oxford University Press, 2005.
Bogen, J. E., D. H. Schultz, and P. J. Vogel. "Completeness of Callosotomy Shown by Magnetic Resonance Imaging in the Long Term." *Archives of Neurology* 45 (1988): 1203–1205.
Bogen, J. E., P. J. Vogel. "Cerebral Commissurotomy in Man." *Bulletin of the Los Angeles Neurological Society* 1962: 27.
Botvinick, M., and J. Cohen. "Rubber Hands 'Feel' Touch That Eyes See." *Nature* 391 (1998): 756.
Corballis, P. M. "Visuospatial processing and the right-hemisphere interpreter." *Brain and Cognition* 53 (2003): 171–176.
Corballis, P. M., M. G. Funnell, and M. S. Gazzaniga. "An Evolutionary Perspective on Hemispheric Asymmetries." *Brain and Cognition* 43 (2000):112–117.
Damasio H, Grabowski T, Frank R, Galaburda AM, Damasio AR. The return of Phineas Gage: clues about the brain from the skull of a famous patient. Science 1994;264:1102-1105.
Devinsky, O. "Delusional Misidentifications and Duplications: Right Brain Lesions, Left Brain Delusions." *Neurology* 72 (2009): 80–87.
Devos, T., and M. R. Banaji. "Implicit Self and Identity." *Annals of the New York Academy of Sciences* 1001 (2003): 177–211.
Espinosa, P. S., C. D. Smith, and J. R. Berger. "Alien Hand Syndrome." *Neurology* 67 (2006): E21.
Feinberg, T. E., and J. P. Keenan. "Where in the Brain Is the Self?" *Conscious and Cognition* 14 (2005): 661–678.

Finger, S. *Origins of Neuroscience: A History of Explorations into Brain Function.* New York: Oxford University Press, 1994.

Flor, H., L. Nikolajsen, and T. Staehelin Jensen. "Phantom Limb Pain: A Case of Maladaptive CNS Plasticity?" *Nature Reviews Neuroscience* 7 (2006): 873–881.

Fontenrose, J. *The Delphic Oracle : Its Responses and Operations with a Catalogue of Responses.* Berkeley, Calif.: University of California Press, 1978.

Gazzaniga, M. S. "Cerebral Specialization and Interhemispheric Communication: Does the Corpus Callosum Enable the Human Condition?" *Brain* 123 (2000) Pt. 7: 1293–1326.

Gazzaniga, M. S. "Forty-Five Years of Split-Brain Research and Still Going Strong." *Nature Reviews Neuroscience* 6 (2005): 653–659.

Gazzaniga, M. S., and J. LeDoux. *The Integrated Mind.* New York: Plenum Press, 1978.

Geschwind, N. "Disconnexion Syndromes in Animals and Man" I. *Brain* 88 (1965): 237–294.

Geschwind, N. "Disconnexion Syndromes in Animals and Man" II. *Brain* 88 (1965): 585–644.

Greenberg, D. B., F. H. Hochberg, and G. B. Murray. "The Theme of Death in Complex Partial Seizures." *American Journal of Psychiatry* 141 (1984): 1587–1589.

Halpern, M. E., J. O. Liang, and J. T. Gamse. "Leaning to the Left: Laterality in the Zebrafish Forebrain." *Trends in Neurosciences* 26 (2003): 308–313.

Immordino-Yang, M. H., A. McColl, H. Damasio, and A. Damasio. "Neural Correlates of Admiration and Compassion." *Proceedings of the National Academy of Science* 106 (2009): 8021–8026.

James, W. *The Principles of Psychology.* New York: Cosimo, 1890.

Janik, V. M., L. S. Sayigh, and R. S. Wells. "Signature Whistle Shape Conveys Identity Information to Bottlenose Dolphins." *Proceedings of the National Academy of Science* 103 (2006): 8293–8297.

Johnson, S. C., L. C. Baxter, L. S. Wilder, J. G. Pipe, J. E. Heiserman, and G. P. Prigatano. "Neural Correlates of Self-Reflection." *Brain* 125 (2002): 1808–1814.

Kandel, E. R. *In Search of Memory: The Emergence of a New Science of Mind,* 1st ed. New York: W. W. Norton & Company, 2006.

Kaufmann, W. A. *Nietzsche: Philosopher, Psychologist, Antichrist,* 3rd ed. Princeton, N.J.: Princeton University Press, 1968.

Klingler, J., and P. Gloor. "The Connections of the Amygdala and of the Anterior Temporal Cortex in the Human Brain." *Journal of Comparative Neurology* 115 (1960): 333–369.

Leary, M. R., and J. P. Tangney. *Handbook of Self and Identity.* New York: Guilford Press, 2003.

LeDoux, J. E. "Synaptic Self : How Our Brains Become Who We Are." New York: Viking, 2002.

LeDoux, J. E., G. L. Risse, S. P. Springer, D. H. Wilson, and M. S. Gazzaniga. "Cognition and Commissurotomy." *Brain* 100 (1977) Pt. 1: 87–104.

Lotze, M., H. Flor, W. Grodd, W. Larbig, and N. Birbaumer. "Phantom Movements and Pain. An fMRI Study in Upper Limb Amputees." *Brain* 124 (2001): 2268–2277.

Luck, S. J., S. A. Hillyard, G. R. Mangun, and M. S. Gazzaniga. "Independent Hemispheric Attentional Systems Mediate Visual Search in Split-Brain Patients." *Nature* 342 (1989): 543–545.

Meares, R. "The Contribution of Hughlings Jackson to an Understanding of Dissociation." *American Journal of Psychiatry* 156 (1999): 1850–1855.

Metzinger, T. *Being No One : The Self-Model Theory of Subjectivity.* Cambridge, Mass.: MIT Press, 2003.

Miller, B.L., W. W. Seeley, P. Mychack, H. J. Rosen, I. Mena, and K. Boone. "Neuroanatomy of the Self: Evidence from Patients with Frontotemporal Dementia." *Neurology* 57 (2001): 817–821.

Morin, A. "Self-Awareness and the Left Hemisphere: The Dark Side of Selectively Reviewing the Literature." *Cortex* 43 (2007): 1068–1073; discussion 1074–1082.

Nagel, T. "What Is Like to Bea Bat?" *Philosophical Review* 83 (1974): 435–450.

The New Oxford Dictionary of English. Oxford, UK: Oxford University Press, 1998.

Nietzsche, F. W. translated by W. A. Kaufmann. *The Portable Nietzsche.* New York: Viking Press, 1968.

Nietzsche, F. W., translated by W. A. Kaufmann. *Beyond Good and Evil: Prelude to a Philosophy of the Future.* New York: Vintage Books, 1966.

Pearn, J., and C. Gardner-Thorpe. "Jules Cotard (1840–1889): His Life and the Unique Syndrome Which Bears His Name." *Neurology* 58 (2002): 1400–1403.

Platek, S. M., K. Wathne, N. G. Tierney, and J. W. Thomson. "Neural Correlates of Self-Face Recognition: An Effect-Location Meta-Analysis." *Brain Research* 1232 (2008): 173–184.

Plotnik, J. M., F. B. de Waal, and D. Reiss. "Self-Recognition in an Asian Elephant." *Proceedings of the National Academy of Science* 103 (2006): 17053–17057.

Ramachandran, V. S., and S. Blakeslee. "*Phantoms in the Brain: Human Nature and the Architecture of the Mind.* London: Fourth Estate, 1998.

Reiss, D., and L. Marino. "Mirror Self-Recognition in the Bottlenose Dolphin: A Case of Cognitive Convergence." *Proceedings of the National Academy of Science* 98 (2001): 5937–5942.

Roser, M. E., J. A. Fugelsang, K. N. Dunbar, P. M. Corballis, and M. S, Gazzaniga. "Dissociating Processes Supporting Causal Perception and Causal Inference in the Brain." *Neuropsychology* 19 (2005): 591–602.

Stace, W. T. *Mysticism and Philosophy.* London: Macmillan, 1960.

Stuss, D. T., G. G. Gallup Jr., and M. P. Alexander. "The Frontal Lobes Are Necessary for 'Theory Of Mind.'" *Brain* 124 (2001): 279–286.

Susic, V., and R. Kovacevic. "Sleep Patterns in Chronic Split-Brain Cats." *Brain Research* 65 (1974): 427–441.

Tanaka, H., M. Arai, T. Kadowaki, H. Takekawa, N. Kokubun, and K. Hirata "Phantom Arm And Leg After Pontine Hemorrhage." *Neurology* 70 (2008): 82–83.

Trujillano, A. C. "Jules Cotard (1840–1889)." *Neurology* 60 (2003): 153; author reply 153.

Wilson, D. H., A. Reeves, and M. Gazzaniga. "Division of the Corpus Callosum for Uncontrollable Epilepsy." *Neurology* 28 (1978): 649–653.

Wolford, G., M. B. Miller, and M. Gazzaniga. "The Left Hemisphere's Role in Hypothesis Formation." *Journal of Neuroscience* 20 (2000): RC64.

Yang, T. T., C. Gallen, B. Schwartz, F. E. Bloom, V. S. Ramachandran, and S. Cobb. "Sensory Maps in the Human Brain." *Nature* 368 (1994): 592–593.

CHAPTER 4

The Greyson near-death experience scale is used to identify and compare near-death experiences (NDE). There are sixteen questions in four categories. Each question is assigned a value of 0 to 2. A maximum possible score is 32, with the more features that happen during the near-death experience, the higher the total score. A total score of 7 is the minimum necessary to consider an event a near-death experience. The principal weakness of the Greyson scale is that the scale is not based on known brain physiology.

Table 8: Greyson scale.

Question	Response
Cognitive (Thoughts)	
1. Did time seem to speed up?	2 = Everything seemed to be happening all at once
	1 = Time seemed to go faster
	0 = Neither
2. Were your thoughts speeded up?	2 = Incredibly fast
	1 = Faster than usual
	0 = Neither
3. Did scenes from your past come back to you?	2 = Past flashed before me, out of my control
	1 = Remembered many past events
	0 = Neither
4. Did you suddenly seem to understand everything?	2 = About the universe
	1 = About myself or others
	0 = Neither
Affective (Feelings)	
5. Did you have a feeling of peace or pleasantness?	2 = Incredible peace or pleasantness
	1 = Relief or calmness
	0 = Neither
6. Did you have a feeling of joy?	2 = Incredible Joy
	1 = Happiness
	0 = Neither
7. Did you feel a sense of harmony or unity with the universe?	2 = United, one with the world
	1 = No longer in conflict with nature
	0 = Neither
8. Did you see or feel surrounded by a brilliant light?	2 = Light clearly of mystical or otherworldly origin

	1 = Unusually bright light 0 = Neither
Paranormal	
9. Were your senses more vivid than usual?	2 = Incredibly more so 1 = More so than usual 0 = Neither
10. Did you seem to be aware of things going on elsewhere, as if by ESP?	2 = Yes, and facts later corroborated 1 = Yes, but facts not yet corroborated 0 = Neither
11. Did scenes from the future come to you?	2 = From the world's future 1 = From personal future 0 = Neither
12. Did you feel separated from your physical body?	2 = Clearly left the body and existed outside it 1 = Lost awareness of the body 0 = Neither
Transcendental	
13. Did you seem to enter some other, unearthly world?	2 = Clearly mystical or unearthly realm 1 = Unfamiliar, strange place 0 = Neither
14. Did you seem to encounter a mystical being or presence?	2 = Definite being, or voice clearly of mystical or otherworldly origin 1 = Unidentifiable voice 0 = Neither
15. Did you see deceased spirits or religious figures?	2 = Saw them 1 = Sensed their presence 0 = Neither
16. Did you come to a border or point of no return?	2 = A barrier I was not permitted to cross; or "sent back" to life involuntarily 1 = A conscious decision to "return" to life 0 = Neither

Ayer, A. J. *The Meaning of Life and Other Essays.* London: Weidenfeld and Nicolson, 1990.

Ayer, A. J. *The Origins of Pragmatism: Studies in the Philosophy of Charles Sanders Peirce and William James.* San Francisco: Freeman, 1968.

Ayer, A. J. "What I Saw When I Was Dead . . ." *Sunday Telegraph,* August 28, 1988.

Blackmore, S. J. *Dying to Live : Near-Death Experiences.* Buffalo, N.Y.: Prometheus Books, 1993.

Busey, G. Available at: http://transcripts.cnn.com/Transcripts/0505/23/lkl.01.html.

"Eyewitness: How Accurate is Visual Memory?" *60 Minutes.* March 8, 2009.

Foges, P. "An Atheist Meets the Masters of the Universe." Available at: http://www.laphamsquarterly.org/roundtable/roundtable/an-atheist-meets-the-masters-of-the-universe.php.

French, C. C. "Dying to Know the Truth: Visions of a Dying Brain, or False Memories?" *Lancet* 358 (2001): 2010–2011.

French, C. C. "Fantastic Memories." *Journal of Consciousness Studies* (2003): 10.

Gallup, G., and W. Proctor. *Adventures in Immortality.* New York: McGraw-Hill, 1982.

Greyson, B. "Incidence and Correlates of Near-Death Experiences in a Cardiac Care Unit." *General Hospital Psychiatry* 25 (2003): 269–276.

Greyson, B. "The Near-Death Experience Scale. Construction, Reliability, and Validity." *Journal of Nervous and Mental Diseases* 171 (1983): 369–375.

Grossberg J. "Roy Recounts Tiger Mauling." Available at: http://video.eonline.com/uberblog/b48260_roy_recounts_tiger_mauling.html.

Hume, D. *Of Miracles.* La Salle, Ill.: Open Court, 1985.

James, W., M. E. Marty. *The Varieties of Religious Experience: A Study In Human Nature.* Harmondsworth, Middlesex, England; New York: Penguin Books, 1982

Jung, C. G., translated by A. Jaffé. *Memories, Dreams, Reflections,* rev. ed. New York: Vintage Books, 1989.

Kellehear, A. *Experiences Near Death: Beyond Medicine and Religion.* New York: Oxford University Press, 1996.

Moody, R. *Elvis After Life.* Atlanta, Ga.: Peachtree Publishers, Ltd., 1987.

Morse, M. L. "Near-Death Experiences of Children." *Journal of Pediatric Oncology Nursing* 11 (1994): 139–144; discussion 145.

Morse, M. L. "Near-death Experiences and Death-Related Visions in Children: Implications for the Clinician." *Current Problems in Pediatrics* 24 (1994): 55–83.

Morse, M., P. Castillo, D. Venecia, J. Milstein, and D. C. Tyler. "Childhood Near-Death Experiences." *American Journal of Diseases of Children* 140 (1986): 1110–1114.

Nelson, K. R., M. Mattingly, S. A. Lee, and F. A. Schmitt. Does the Arousal System Contribute to Near-death Experience?" *Neurology* 66 (2006): 1003–1009.

Pasricha, S., and I. Stevenson. "Near-Death Experiences in India. A Preliminary Report." *Journal of Nervous and Mental Disease* 174 (1986): 165–170.

Petito, C. K., E. Feldmann, W. A. Pulsinelli, and F. Plum. "Delayed Hippocampal Damage in Humans Following Cardiorespiratory Arrest." *Neurology* 37 (1987): 1281–1286.

Plato, H. D. P. Lee. *The Republic*, 2nd ed. (revised). Harmondsworth, England; Baltimore: Penguin, 1987.

The Innocence Project. Available at: http://www.innocenceproject.org/.

Russell, B. *Mysticism and Logic*. Mineola, N.Y.: Dover Publications, 2004.

Seymour, J., May 23, 2005. Available at: http://transcripts.cnn.com/Transcripts/0505/23/lkl.01.html.

Stone, S. Available at: http://transcripts.cnn.com/Transcripts/0505/23/lkl.01.html.

Symonds, J. A., and H. F. Brown. *John Addington Symonds: A Biography*. London: J. C. Nimmo, 1895.

Taylor, E. "Life=Passion." *America's AIDS Magazine*, February 2003.

Thompson-Cannino, J. *Picking Cotton : Our Memoir of Injustice and Redemption*. New York: St. Martin's Press, 2009.

Walker, F. O. "A Nowhere Near-Death Experience: Heavenly Choirs Interrupt Myelography." *Journal of the American Medical Association* 261 (1989): 3245–3246.

Wilhelm, R., and C. F. Baynes. *The I Ching or Book of Changes*, 3rd. ed. Princeton, N.J.: Princeton University Press for the Bollingen Foundation, 1967.

Yamamura, H. [Implication of near-death experience for the elderly in terminal care]. *Nippon Ronen Igakkai Zasshi* 35 (1998): 103–115.

CHAPTER 5

Barrera-Mera, B., and E. Barrera-Calva. "The Cartesian Clock Metaphor for Pineal Gland Operation Pervades the Origin of Modern Chronobiology." *Neuroscience and Biobehavioral Reviews* 23 (1998): 1–4.

Benson, A. J. "Spatial Disorientation—Common Illusions." In: Ernsting J., Nicholson A.N., and D. J. Rainford, ed. *Aviation Medicine*, 3rd ed. Oxford: Butterworth & Heinmann, 1999, pp. 437–454.

Blackmore, S. "Out-of-Body Experiences in Schizophrenia. A Questionnaire Survey." *Journal of Nervous and Mental Disease* 174 (1986): 615–619.

Blackmore, S. J. *Beyond the Body: An Investigation of Out-of-the-Body Experiences*. Chicago: Academy Chicago Publishers, 1992.

Blanke, O., T. Landis, L. Spinelli, and M. Seeck. "Out-of-Body Experience and Autoscopy of Neurological Origin." *Brain* 127 (2004): 243–258.

Blanke, O., S. Ortigue, T. Landis, and M. Seeck. "Stimulating Illusory Own-Body Perceptions." *Nature* 419 (2002): 269–270.

Brugger, P., M. Regard, T. Landis, and O. Oelz. "Hallucinatory Experiences in Extreme-Altitude Climbers." *Neuropsychiatry, Neuropsychology, and Behavioral Neurology* 12 (1999): 67–71.

Cobcroft, M. D., and C. Forsdick. "Awareness Under Anaesthesia: The Patients' Point of View." *Anaesthesia and Intensive Care* 21 (1993): 837–843.

Damasio, A. R. "Descartes' Error : Emotion, Reason, and the Human Brain." New York: Avon Books, 1995.

Descartes, R., J. Cottingham, and B. A. O. Williams. *Meditations on First Philosophy: With Selections from the Objections and Replies*, rev. ed. New York: Cambridge University Press, 1996.

Devinsky, O., E. Feldmann, K. Burrowes, and E. Bromfield. "Autoscopic Phenomena with Seizures." *Archives of Neurology* 46 (1989): 1080–1088.

Finger, S. "Descartes and the Pineal Gland in Animals: A Frequent Misinterpretation." *Journal of the History of the Neurosciences* 4 (1995): 166–182.

Firth, P. G., and H. Bolay. "Transient High Altitude Neurological Dysfunction: An Origin in the Temporoparietal Cortex." *High Altitude Medicine and Biology* 5 (2004): 71–75.

Greyson, B. "Incidence and Correlates of Near-Death Experiences in a Cardiac Care Unit." *General Hospital Psychiatry* 25 (2003): 269–276.

Gupta, S. *Cheating Death: The Doctors and Medical Miracles That Are Saving Lives Against All Odds*. New York: Hachette Book Group, 2009.

Harper, C. M., and K. R. Nelson. "Intraoperative Electrophysiological Monitoring in Children." *Journal of Clinical Neurophysiology* 9 (1992): 342–356.

Heimer, L., and G. W. Van Hoesen. "The Limbic Lobe and Its Output Channels: Implications for Emotional Functions and Adaptive Behavior." *Neuroscience & Biobehavioral Reviews* 30 (2006): 126–147.

Horstmann, A., S. Frisch, R. T. Jentzsch, K. Muller, A. Villringer, and M. L. Schroeter. "Resuscitating the Heart But Losing the Brain: Brain Atrophy in the Aftermath of Cardiac Arrest." *Neurology* 74 (2010): 306–312.

Hotchkiss, R. S., A. Strasser, J. E. McDunn, and P. E. Swanson. "Cell Death." *New England Journal of Medicine* 361 (2009): 1570–1583.

Jackson, D. A. "Out-of-Body Experience in a Patient Emerging from Anesthesia." *Journal of Post Anesthesia Nursing* 10 (1995): 27–28.

Jansen, K. L. "Neuroscience and the Near-Death Experience: Roles for the NMSA-PCP Receptor, the Sigma Receptor and the Endopsychosins." *Medical Hypotheses* 31 (1990): 25–29.

Lambert, E. H., and E. H.Wood. "The Problem of Blackout and Unconsciousness in Aviators." *Medical Clinics of North America.* 30 (1946): 833–844.

Lempert, T., M. Bauer, and D. Schmidt. "Syncope: A Videometric Analysis of 56 Episodes of Transient Cerebral Hypoxia." *Annals of Neurology.* 36 (1994): 233–237.

Lempert, T., M. Bauer, and D. Schmidt. "Syncope and Near-Death Experience." *Lancet* 344 (1994): 829–830.

Lokhorst, G. J., and T. T. Kaitaro. "The Originality of Descartes' Theory About the Pineal Gland." *Journal of the History of the Neurosciences* 10 (2001): 6–18.

Mano, H., and Y. Fukada. " 'A Median Third Eye' Pineal Gland Retraces Evolution of Vertebrate Photoreceptive Organs." *Photochemistry and Photobiology.* phot.allenpress.com 2006:DOI: 10.1562/2006-1502-1524-IR-1813.

Maronde, E., and J. H. Stehle. "The Mammalian Pineal Gland: Known Facts, Unknown Facets." *Trends in Endocrinology and Metabolism* 18 (2007): 142–149.

Mitchell, J. P. "Activity in Right Temporo-Parietal Junction Is Not Selective for Theory-of-Mind." *Cerebral Cortex* 18 (2008): 262–271.

Moody, R. *Life After Life.* Covington, GA: Mockingbird Books, 1975.

Morse, M., P. Castillo, D. Venecia, J. Milstein, and D. C. Tyler. "Childhood Near-Death Experiences." *American Journal of Diseases of Children* 140 (1986): 1110–1114.

Near Death Research Foundation (NDER). Available at: http://www.nderf.org/polls.htm.

Ohayon, M. M. "Prevalence of Hallucinations and Their Pathological Associations in the General Population." *Journal of Psychiatry Research* 97 (2000): 153–164.

Owens, J. E., E. W. Cook, and I. Stevenson. "Features of 'Near-Death Experience' in Relation to Whether or Not Patients Were Near Death." *Lancet* 336 (1990): 1175–1177.

Pal, H. R., N. Berry, R. Kumar, and R. Ray. "Ketamine Dependence." *Anaesthesia and Intensive Care* 30 (2002): 382–384.

Parker-Pope, T. " 'Choking' Game Deaths on the Rise." Available at: http://well.blogs.nytimes.com/2008/02/14/choking-game-deaths-on-the-rise/.

Parnia, S., D. G. Waller, R. Yeates, and P. Fenwick P. "A Qualitative and Quantitative Study of the Incidence, Features and Aetiology of Near-Death Experiences in Cardiac Arrest Survivors." *Resuscitation* 48 (2001): 149–156.

Penfield, W. "Ferrier Lecture." *Proceedings of the Royal Society of London* 134 (1947): 329–347.

Penfield, W. "The Twenty-ninth Maudsley Lecture: The Role of the Temporal Cortex in Certain Psychical Phenomena." *Journal of Mental Science* 101 (1955): 451–465.

Podoll, K, and D. Robinson . "Out-of-Body Experiences and Related Phenomena in Migraine Art." *Cephalalgia* 19 (1999): 886–896.

Posner, J. B., C. B. Saper, N. D. Schiff, F. Plum. *Plum and Posner's Diagnosis of Stupor and Coma*. In: *Contemporary Neurology Series* 71, 4th ed. Oxford and New York: Oxford University Press, 2007.

Rohricht, F., and S. Priebe. "Disturbances of Body Experience in Schizophrenic Patients." *Fortschritte der Neurologie-Psychiatrie* 65 (1997): 323–336.

Ruby, P., and J. Decety. "Effect of Subjective Perspective Taking During Simulation of Action: A PET Investigation of Agency." *Nature Neuroscience* 4 (2001): 546–550.

Ruby, P., and J. Decety. "How Would You Feel Versus How Do You Think She Would Feel? A Neuroimaging Study of Perspective-Taking with Social Emotions." *Journal of Cognitive Neuroscience* 16 (2004): 988–999.

Sabom, M. B. *Light and Death*. Grand Rapids, Mich.: Zondervan Publishing House, 1998.

Sagan C. *Broca's Brain : Reflections on the Romance of Science*, 1st ed. New York: Random House, 1979.

Sandin, R. H., G. Enlund, P. Samuelsson, and C. Lennmarken. "Awareness During Anaesthesia: A Prospective Case Study." *Lancet* 355 (2000): 707–711.

Saxe, R., and A. Wexler. "Making Sense of Another Mind: The Role of the Right Temporo-Parietal Junction." *Neuropsychologia* 43 (2005): 1391–1399.

Schnipper, J. L., and W. N. Kapoor. "Diagnostic Evaluation and Management of Patients with Syncope." *Medical Clinics of North America* 85 (2001): 423–456, xi.

Smith, C. U. "Descartes' Pineal Neuropsychology." *Brain and Cognition* 36 (1998): 57–72.

Van Lommel, P., R. van Wees, V. Meyers, and I. Elfferich. "Near-Death Experience in Survivors of Cardiac Arrest: A Prospective Study in the Netherlands." *Lancet* 358 (2001):2039–2045.

CHAPTER 6

Aston-Jones G., J. Rajkowski, and J. Cohen. "Locus Coeruleus and Regulation of Behavioral Flexibility and Attention." *Progress in Brain Research* 126 (2000): 165–182.

Aston-Jones, G., J. Rajkowski, and J. Cohen. "Role of Locus Coeruleus in Attention and Behavioral Flexibility." *Biological Psychiatry* 46 (1999): 1309–1320.

Balter, M. "Did *Homo erectus* Tame Fire First?" *Science* 268 (1995): 1570.

Benison, S., A. C. Barger, and E. L. Wolfe. *Walter B. Cannon: The Life and Times of a Young Scientist.* Cambridge, Mass.: Belknap Press, 1987.

Bradford Cannon Papers 1923–2003 H MS c240 Harvard Medical Library, Francis A. Countway Library of Medicine, Boston, Mass.

Brewin, C. R. "What Is It That a Neurobiological Model of PTSD Must Explain?" *Progress in Brain Research* 167 (2008): 217–228.

Cevik, C., M. Otahbachi, E. Miller, S. Bagdure, and K. M. Nugent. "Acute Stress Cardiomyopathy and Deaths Associated with Electronic Weapons." *International Journal of Cardiology* 132 (2009): 312–317.

Corbetta, M., G. Patel, and G. L. Shulman. "The Reorienting System of the Human Brain: From Environment to Theory of Mind." *Neuron* 58 (2008): 306–324.

Damasio, A. R. "Descartes' Error : Emotion, Reason, and the Human Brain." New York: Avon Books, 1995.

Darwin, C., and N. Barlow. *The Autobiography of Charles Darwin, 1809–1882.* New York: Norton & Co., 1969.

Dostoyevsky, F., trans. C. Garnett. *The Idiot.* New York: Dell, 1959.

Eagleman, D. M. "Human Time Perception and Its Illusions." *Current Opinion in Neurobiology* 18 (2008): 131–136.

Frank, J. *Dostoevsky, the Years of Ordeal, 1850–1859.* Princeton, N.J.: Princeton University Press, 1983.

The Free Fall Research Page. Available at: http://www.greenharbor.com/fffolder/ffresearch.html.

Grabenhorst, F., E. T. Rolls, and A. Bilderbeck. "How Cognition Modulates Affective Responses to Taste and Flavor: Top-Down Influences on the Orbitofrontal and Pregenual Cingulate Cortices." *Cerebral Cortex* 18 (2008): 1549–1559.

Heims, H. C., H. D. Critchley, R. Dolan, C. J. Mathias, and L. Cipolotti. "Social and Motivational Function Is Not Critically Dependent on Feedback of Autonomic Responses: Neuropsychological Evidence from Patients with Pure Autonomic Failure." *Neuropsychologia* 42 (2004): 1979–1988.

James, W. "The Physical Basis of Emotion." *Psychological Review* 1 (1894): 516–529.

Jung, C. G., translated by Jaffé A. *Memories, Dreams, Reflections*, rev. ed. New York: Vintage Books, 1989.

Leakey, R. E. *The Origin of Humankind.* New York: BasicBooks, 1994.

McGaugh, J. L. "The Amygdala Modulates the Consolidation of Memories of Emotionally Arousing Experiences." *Annual Review of Neuroscience* 27 (2004): 1–28.

Price, J. L. "Free Will Versus Survival: Brain Systems That Underlie Intrinsic Constraints on Behavior." *Journal of Comparative Neurology* 493 (2005): 132–139.

Rolls, E. T., and C. McCabe. "Enhanced Affective Brain Representations of Chocolate in Cravers vs. Non-Cravers." *European Journal of Neuroscience* 26 (2007): 1067–1076.

Samuels, M. A. "'Voodoo' Death Revisited: The Modern Lessons of Neurocardiology." *Cleveland Clinic Journal of Medicine* 74 (2007) Suppl 1: S8–16.

Sara, S. J. "The Locus Coeruleus and Noradrenergic Modulation of Cognition." *Nature Reviews Neuroscience* 10 (2009): 211–223.

Stetson, C., M. P. Fiesta, and D. M. Eagleman. "Does Time Really Slow Down During a Frightening Event?" PLoS ONE (2007): e1295.

Tsuchiya, N., F. Moradi, C. Felsen, M. Yamazaki, and R. Adolphs. "Intact Rapid Detection of Fearful Faces in the Absence of the Amygdala." *Nature Neuroscience* 2 (2009): 1224–1225.

Usher, M., J. D. Cohen, D. Servan-Schreiber, J. Rajkowski, and G. Aston-Jones. "The Role of Locus Coeruleus in the Regulation of Cognitive Performance." *Science* 283 (1999): 549–554.

Wolfe, E. L., A. C. Barger, and S. Benison. *Walter B. Cannon, Science and Society*. Cambridge, Mass.: Boston Medical Library in the Francis A. Countway Library of Medicine and distributed by the Harvard University Press, 2000.

CHAPTER 7

Alderson, H. L., V. J. Brown, M. P. Latimer, P. J. Brasted, A. H. Robertson, and P. Winn. "The Effect of Excitotoxic Lesions of the Pedunculopontine Tegmental Nucleus on Performance of a Progressive Ratio Schedule of Reinforcement." *Neuroscience* 112 (2002): 417–425.

Aldrich, M. S. "The Clinical Spectrum of Narcolepsy and Idiopathic Hypersomnia." *Neurology* 46 (1996): 393–401.

Arnulf, I., A. M. Bonnet, and P. Damier, et al. "Hallucinations, REM Sleep, and Parkinson's Disease: A Medical Hypothesis." *Neurology* 55 (2000): 281–288.

Augustine, Chadwick H. (translator). *Confessions*. Oxford: Oxford University Press, 1991.

Bandler, R., K. A. Keay, N. Floyd, and J. Price. "Central Circuits Mediating Patterned Autonomic Activity During Active vs. Passive Emotional Coping." *Brain Research Bulletin* 53: 95–104.

Benson, A. J. "Spatial Disorientation-Common Illusions." In: Ernsting, J. Nicholson A.N. , and D. J. Rainford, ed. *Aviation Medicine*, 3rd ed. Oxford: Butterworth & Heinmann, 1999, pp. 437–454.

Blackmore, S. J. *Beyond the Body: An Investigation of Out-of-the-Body Experiences*. Chicago: Academy Chicago Publishers, 1992.

Bootzin, R. R., J. F. Kihlstrom, and D. L. Schacter. *Sleep and Cognition*, 1st ed. Washington, D.C.: American Psychological Association, 1990.

Braun, A. R., T. J. Balkin, N. J. Wesensten, F. Gwadry, R. E. Carson, M. Varga, P. Baldwin, et al. "Dissociated Pattern of Activity in Visual Cortices and Their Projections During Human Rapid Eye Movement Sleep." *Science* 279 (1998): 91–95.

Braun, A. R., T. J. Balkin, N. J. Wesenten, R. E. Carson, M. Varga, P. Baldwin, S. Selbie, et al. "Regional Cerebral Blood Flow Throughout the Sleep-Wake Cycle. An H2(15)O PET Study." *Brain* 120 (1997): 1173–1197.

Broughton, R., V. Valley, M. Aguirre, J. Roberts, W. Suwalski, and W. Dunham. "Excessive Daytime Sleepiness and the Pathophysiology of Narcolepsy-Cataplexy: A Laboratory Perspective." *Sleep* 9 (1986): 205–215.

Buzzi, G. "Near-Death Experiences." *Lancet* 359 (2002): 2116–2117.

Calvo, J. M., S. Datta, J. Quattrochi, and J. A. Hobson. "Cholinergic Microstimulation of the Peribrachial Nucleus in the Cat. II. Delayed and Prolonged Increases in REM Sleep." *Archives of Italian Biology* 130 (1992): 285–301.

Cami, J., and M. Farre. "Drug Addiction." *New England Journal of Medicine* 349 (2003): 975–986.

Cheyne, J. A., and T. A. Girard. "The Body Unbound: Vestibular-Motor Hallucinations and Out-of-Body Experiences." *Cortex* 45 (2009): 201–215.

Cheyne, J. A., S. D. Rueffer, and I. R. Newby-Clark. "Hypnagogic and Hypnopompic Hallucinations During Sleep Paralysis: Neurological and Cultural Construction of the Night-Mare." *Consciousness and Cognition* 8 (1999): 319–337.

Cicogna, P. C., and M. Bosinelli. "Consciousness During Dreams." *Conscious and Cognition* 10 (2001): 26–41.

Cochen, V., I. Arnulf, and S. Demeret, M. L. Neulat, V. Gourlet, X. Drouot, S. Moutereau, et al. "Vivid Dreams, Hallucinations, Psychosis and REM Sleep in Guillain-Barré Syndrome." *Brain* 128 (2005): 2535–2545.

Dahan, L., B. Astier, N. Vautrelle, N. Urbain, B. Kocsis, and G. Chouvet. "Prominent Burst Firing of Dopaminergic Neurons in the Ventral Tegmental Area During Paradoxical Sleep." *Neuropsychopharmacology* 32 (2007): 1232–1241.

Datta, S., and J. A. Hobson. "Neuronal Activity in the Caudolateral Peribrachial Pons: Relationship to PGO Waves and Rapid Eye Movements." *Journal of Neurophysiology* 71 (1994): 95–109.

Datta, S., and J. A. Hobson. "Suppression of Ponto-Geniculo-Occipital Waves by Neurotoxic Lesions of Pontine Caudo-Lateral Peribrachial Cells." *Neuroscience* 67 (1995): 703–712.

Datta, S., E. H. Patterson, and D. F. Siwek. "Brainstem Afferents of the Cholinoceptive Pontine Wave Generation Sites in the Rat." *Sleep Research Online* 2 (1999): 79–82.

Datta, S., J. M. Calvo, J. Quattrochi, and J. A. Hobson. "Cholinergic Microstimulation of the Peribrachial Nucleus in the Cat. I. Immediate and Prolonged Increases in Ponto-Geniculo-Occipital Waves." *Archives of Italian Biology* 130 (1992): 263–284.

Fernandez-Guardiola, A., A. Martinez, A. Valdes-Cruz, V. M. Magdaleno-Madrigal, D. Martinez, and R. Fernandez-Mas. "Vagus Nerve Prolonged Stimulation in Cats: Effects on Epileptogenesis (Amygdala Electrical Kindling): Behavioral and Electrographic Changes." *Epilepsia* 40 (1999): 822–829.

Forster, E. M., and J. E. Whinnery. "Recovery from Gz-Induced Loss of Consciousness: Psychophysiologic Considerations." *Aviation, Space, and Environmental Medicine* 59 (1988): 517–522.

Foutz, A. S., J. P. Ternaux, and J. J. Puizillout. "Les Stades de Sommeil de la Preparation 'Encephale Isole': II. Phases Paradoxales. Leur Declenchement par la Stimulation des Afferences Baroceptives." *Electroencephalography and Clinical Neurophysiology* 37 (1974): 577–588.

French, C. C., J. Santomauro, V. Hamilton, R. Fox, and M. A. Thalbourne. "Psychological Aspects of the Alien Contact Experience." *Cortex* 44 (2008): 1387–1395.

Fukuda, K., A. Miyasita, M. Inugami, and K. Ishihara. "High Prevalence of Isolated Sleep Paralysis: Kanashibari Phenomenon in Japan." *Sleep* 10 (1987): 279–286.

Greyson, B., and N. E. Bush. "Distressing Near-Death Experiences." *Psychiatry* 55 (1992): 95–110.

Greyson, B. "Consistency of Near-Death Experience Accounts Over Two Decades: Are Reports Embellished Over Time?" *Resuscitation* 73 (2007): 407–411.

Greyson, B. "Posttraumatic Stress Symptoms Following Near-Death Experiences." *American Journal of Orthopsychiatry* 71 (2001): 368–373.

Greyson, B. "Varieties of Near-Death Experience." *Psychiatry* 56 (1993): 390–399.

Hishikawa, Y., H. Koida, K. Yoshino, H. Wakamatsu, Y. Sugita, and S. Iijima. "Characteristics of REM Sleep Accompanied by Sleep Paralysis and Hypnagogic Hallucinations in Narcoleptic Patients." *Waking Sleeping* 2 (1978): 113–123.

Hobson, J. A., *Dreaming as Delirium: How the Brain Goes Out of Its Mind*, 1st MIT Press ed. Cambridge, Mass.: MIT Press, 1999.

Hobson, J. A., S. A. Hoffman, R. Helfand, and D. Kostner. "Dream Bizarreness and the Activation-Synthesis Hypothesis." *Human Neurobiology* 6 (1987): 157–164.

Hobson, J. A. "REM Sleep and Dreaming: Towards a Theory of Protoconsciousness." *Nature Reviews Neuroscience* 10 (2009): 803–813.

Holden, K .J., and C. C. French. "Alien Abduction Experiences: Some Clues from Neuropsychology and Neuropsychiatry." *Cognitive Neuropsychiatry* 7 (2002): 163–178.

Kahn, D., E. Pace-Schott, J. A. Hobson. "Emotion and Cognition: Feeling and Character Identification in Dreaming." *Consciousness and Cognition* 11 (2002): 34–50.

Kahn, D., R. Stickgold, E. F. Pace-Schott, and J. A. Hobson. "Dreaming and Waking Consciousness: A Character Recognition Study." *Journal of Sleep Research* 9 (2000): 317–325.

Kaur, S., S. Thankachan, S. Begum, M. Liu, C. Blanco-Centurion, and P. J. Shiromani. "Hypocretin-2 Saporin Lesions of the Ventrolateral Periaquaductal Gray (vlPAG) Increase REM Sleep in Hypocretin Knockout Mice." *PLoS ONE* 4 (2009): e6346.

Keay, K. A., C. I. Clement, W. M. Matar, D. J. Heslop, L. A. Henderson, and R. Bandler. "Noxious Activation of Spinal or Vagal Afferents Evokes Distinct Patterns of Fos-Like Immunoreactivity in the Ventrolateral Periaqueductal Gray of Unanaesthetised Rats." *Brain Research* 948 (2002): 122–130.

Kumar, R., S. Behari, J. Wahi, D. Banerji, and K. Sharma. "Peduncular Hallucinosis: An Unusual Sequel to Surgical Intervention in the Suprasellar Region." *British Journal of Neurosurgery* 13 (1999): 500–503.

LaBerge, S., and D. J. DeGracia. "Vareties of Lucid Dreaming Experience." In: Kunzendorf. R. G., and B. Wallace, eds. *Individual Differences in Conscious Experience*. Amsterdam: John Benjamins, 2000.

LaBerge, S., and H. Rheingold. *Exploring the World of Lucid Dreaming*. New York: Ballantine, 1990.

LaBerge, S., L. E. Nagel, W. C. Dement, and V. P. Zarcone Jr. "Lucid Dreaming Verified by Volitional Communication During REM Sleep." *Perceptual & Motor Skills* 52 (1981): 727–732.

LaBerge, S. *Lucid Dreaming*, 1st ed. Los Angeles, Boston: J.P. Tarcher, 1985.

LaBerge, S. "Lucid Dreaming: Psychophysiological Studies of Consciousness During REM Sleep." In: Bootzin, R. R., J. F. Kihlstrom, and D. L. Schacter, eds. *Sleep and Cognition*, 1st ed. Washington, D.C.: American Psychological Association, 1990: pp. xvii, 205.

LaBerge, S. "Lucid Dreaming as a Learnable Skill: A Case Study." *Perceptual & Motor Skills* 51 (1980): 1039–1042.

LaBerge S., L. Levitan, A. Brylowski, and W. Dement. "'Out-of-Body' Experiences Occurring in REM Sleep." *Sleep Research* 17 (1988): 115.

Lambert, E. H., E. H. Wood. "The Problem of Blackout and Unconsciousness in Aviators." *Medical Clinics of North America* 30 (1946): 833–844.

Laureys, S., G. Tononi, O. Blanke, and S. Dieguez. "Leaving Body and Life Behind: Out-of-Body and Near-Death Experience." *The Neurology of Consciousness: Cognitive Neuroscience and Neuropathology,* 1st ed. Amsterdam, Boston, and London: Academic, 2009 pp. 303–325.

Lu, J., D. Sherman, M. Devor, and C. B. Saper. "A Putative Flip-flop Switch for Control of REM Sleep." *Nature* 441 (2006): 589–594.

Mahowald, M. W., and C. H. Schenck. "Dissociated States of Wakefulness and Sleep." *Neurology* 42 (1992): 44–51.

Mahowald, M. W. "What State Dissociation Can Teach Us About Consciousness and the Function of Sleep." *Sleep Medicine* 10 (2009): 159–160.

Malow, B. A., J. Edwards, M. Marzec,O. Sagher, D. Ross, and G. Fromes. "Vagus Nerve Stimulation Reduces Daytime Sleepiness in Epilepsy Patients." *Neurology* 57 (2001): 879–884.

Manford, M., and F. Andermann. "Complex Visual Hallucinations. Clinical and Neurobiological Insights." *Brain* 121 (1998): 1819–1840.

Maquet, P., P. Ruby, A. Maudoux, G.Albouy, V. Sterpenich, T. Dang-Vu, M. Desseilles, et al. "Human Cognition During REM Sleep and the Activity Profile Within Frontal and Parietal Cortices: A Reappraisal of Functional Neuroimaging Data." *Progress in Brain Research* 150 (2005): 219–227.

McCarley, R. W., and E. Hoffman. "REM Sleep Dreams and the Activation-Synthesis Hypothesis." *American Journal of Psychiatry* 138 (1981): 904–912.

McCarley, R. W., O. Benoit, and G. Barrionuevo. "Lateral Geniculate Nucleus Unitary Discharge in Sleep and Waking: State- and Rate-Specific Aspects." *Journal of Neurophysiology* 50 (1983): 798–818.

Merritt, J. M., R. Stickgold, E. Pace-Schott, E. F. Williams, and J. A. Hobson. "Emotion Profiles in the Dreams of Young Adult Men and Women." *Consciousness and Cognition* 3 (1994): 46–60.

Nelson, K. R., M. Mattingly, and F. A. Schmitt. "Out-of-Body Experience and Arousal." *Neurology* 68 (2007): 794–795.

Nelson, K. R., M. Mattingly, S. A. Lee, and F. A. Schmitt. "Does the Arousal System Contribute to Near-Death Experience?" *Neurology* 66 (2006): 1003–1009.

Ness, R. C. "The Old Hag Phenomenon as Sleep Paralysis: A Biocultural Interpretation." *Culture, Medicine and Psychiatry* 2 (1978): 15–39.

Nielsen, T. A. "Mentation During Sleep: The NREM / REM Distinction." In: Lydic, R., and H. A. Baghdoyan, eds. *Handbook of Behavioral State Control: Molecular and Cellular Mechanisms.* Boca Raton, Fla.: CRC Press, 1999, pp. 101–128.

Nofzinger, E. A., M. A. Mintun, M. Wiseman, D. J. Kupfer, and R. Y. Moore. "Forebrain Activation in REM Sleep: An FDG PET Study." *Brain Research* 770 (1997): 192–201.

Noyes, R., Jr., and R. Kletti. "Depersonalization in the Face of Life-Threatening Danger: A Description." *Psychiatry* 39 (1976): 19–27.

Oakman, S. A., P. L. Faris, P. E. Kerr, C. Cozzari, and B. K. Hartman. "Distribution of Pontomesencephalic Cholinergic Neurons Projecting to Substantia Nigra Differs Significantly from Those Projecting to Ventral Tegmental Area." *Journal of Neuroscience* 15 (1995): 5859–5869.

Ohayon, M. M., R. G. Priest, J. Zulley, S. Smirne, and T. Paiva. "Prevalence of Narcolepsy Symptomatology and Diagnosis in the European General Population." *Neurology* 58 (2002): 1826–1833.

Olmstead, M. C., E. M. Munn, K. B. Franklin, and R. A, Wise. "Effects of Pedunculopontine Tegmental Nucleus Lesions on Responding for Intravenous Heroin Under Different Schedules of Reinforcement." *Journal of Neuroscience* 18 (1998): 5035–5044.

Overeem, S., E. Mignot, J. G. van Dijk, and G. J. Lammers. "Narcolepsy: Clinical Features, New Pathophysiologic Insights, and Future Perspectives." *Journal of Clinical Neurophysiology* 18 (2001): 78–105.

Overney, L. S., S. Arzy, and O. Blanke. "Deficient Mental Own-Body Imagery in a Neurological Patient with Out-of-Body Experiences Due to Cannabis Use." *Cortex* 45 (2009): 228–235.

Owens, J. E., E. W. Cook, and I. Stevenson. "Near-Death Experience." *Lancet* 337 (1991): 1167–1168.

Persson, B., and T. H. Svensson. "Control of Behaviour and Brain Noradrenaline Neurons by Peripheral Blood Volume Receptors." *Journal of Neural Transmission* 52 (1981): 73–82.

Puizillout, J. J., and A. S. Foutz. "Characteristics of the Experimental Reflex Sleep Induced by Vago-Aortic Nerve Stimulation." *Electroencephalography and Clinical Neurophysiology* 42 (1977): 552–563.

Puizillout, J. J., and A. S. Foutz. "Vago-Aortic Nerves Stimulation and REM Sleep: Evidence for a REM-Triggering and a REM-Maintenance Factor." *Brain Research* 111 (1976): 181–184.

Rechtschaffen, A., B. M. Bergmann, C. A. Everson, C. A. Kushida, and M. A. Gilliland. "Sleep Deprivation in the Rat: X. Integration and Discussion of the Findings." *Sleep* 12 (1989): 68–87.

Revonsuo, A. "The Reinterpretation of Dreams: An Evolutionary Hypothesis of the Function of Dreaming." *Behavioral and Brain Sciences* 23 (2000): 877–901; discussion 904–1121.

Reynolds, M., and C. R. Brewin. "Intrusive Memories in Depression and Posttraumatic Stress Disorder." *Behavior Research and Therapy* 37 (1999): 201–215.

Ribary, U. "Dynamics of Thalamo-Cortical Network Oscillations and Human Perception." *Progress in Brain Research* 150 (2005): 127–142.

Sabom, M. B. *Recollections of Death: A Medical Investigation*. New York: Harper & Row, 1982.

Saito, H., K. Sakai, and M. Jouvet. "Discharge Patterns of the Nucleus Parabrachialis Lateralis Neurons of the Cat During Sleep and Waking." *Brain Research* 134 (1977): 59–72.

Semba, K., and H. C. Fibiger. "Afferent Connections of the Laterodorsal and the Pedunculopontine Tegmental Nuclei in the Rat: A Retro- and Antero-Grade Transport and Immunohistochemical Study." *Journal of Comparative Neurology* 323 (1992): 387–410.

Siegel, J. M. "The REM Sleep-Memory Consolidation Hypothesis." *Science* 294 (2001): 1058–1063.

Siegel, J. M. "Clues to the Functions of Mammalian Sleep." *Nature* 437 (2005): 1264–1271.

Solms, M. "*The Neuropsychology Of Dreams: A Clinico-Anatomical Study*. Mahwah, N.J.: Erlbaum, 1997.

Stickgold, R., J. A. Hobson, R. Fosse, and M. Fosse. "Sleep, Learning, and Dreams: Off-Line Memory Reprocessing." *Science* 294 (2001): 1052–1057.

Tachibana, M., K. Tanaka, Y. Hishikawa, and Z. Kaneko. "A Sleep Study of Acute Psychotic States Due to Alcohol and Meprobamate Addiction." New York: Spectrum Publications, 1975, pp. 177–205.

Takeuchi, T., A. Miyasita, Y. Sasaki, M. Inugami, and K. Fukuda. "Isolated Sleep Paralysis Elicited by Sleep Interruption." *Sleep* 15 (1992): 217–225.

Tsukamoto, H., T. Matsushima, S. Fujiwara, and M. Fukui. "Peduncular Hallucinosis Following Microvascular Decompression for Trigeminal Neuralgia: Case Report." *Surgical Neurology* 40 (1993): 31–34.

Vagg, D. J., R. Bandler, and K. A. Keay. "Hypovolemic Shock: Critical Involvement of a Projection from the Ventrolateral Periaqueductal Gray to the Caudal Midline Medulla." *Neuroscience* 152 (2008): 1099–1109.

Valdes-Cruz, A., V. M. Magdaleno-Madrigal, D. Martinez-Vargas D, R. Mas-Fernández, S. Almazán-Alvarado, A. Martinez, A. Fernández-Guardiola, et al. "Chronic Stimulation of the Cat Vagus Nerve: Effect on Sleep and Behavior." *Progress jn Neuro-Psychopharmacology and Biological Psychiatry* 26 (2002): 113–118.

Valli, K, and A. Revonsuo. "The Threat Simulation Theory in Light of Recent Empirical Evidence: A Review." *American Journal of Psychology* 122 (2009): 17–38.

Voss, U., R. Holzmann, I. Tuin, and J. A. Hobson. "Lucid Dreaming: A State of Consciousness with Features of Both Waking and Non-Lucid Dreaming." *Sleep* 32 (2009): 1191–1200.

Whinnery, J. E., and A. M. Whinnery. "Acceleration-Induced Loss of Consciousness. A Review of 500 Episodes." *Archives of Neurology*. 47 (1990): 764–776.

Yeomans, J. S., A. Mathur, and M. Tampakeras. "Rewarding Brain Stimulation: Role of Tegmental Cholinergic Neurons That Activate Dopamine Neurons." *Behavioral Neuroscience* 107 (1993): 1077–1087.

CHAPTER 8

Aghajanian, G. K., and G. J. Marek. "Serotonin and Hallucinogens." *Neuropsychopharmacology* 21 (1999): 16S–23S.

Table 9: The qualities of mystical experience based on Stace.

Extrovertive
The unifying vision—all things are One Perceived through the senses
Concrete understanding of the One in all things
Introvertive
Unitary consciousness; the One, "pure consciousness"
Beyond space and time
Both
Sense of objective reality
Blessedness, peace, or joy
Feeling the holy, sacred, or divine has been touched
Paradoxicality
Ineffable*

* Only with reservation does Stace consider extrovertive mystical experience ineffable.

Amat, J., M. V. Baratta, E. Paul, S. T. Bland, L. R. Watkins, and S. F. Maier. "Medial Prefrontal Cortex Determines How Stressor Controllability Affects Behavior and Dorsal Raphe Nucleus." *Nature Neuroscience* 8 (2005): 365–371.

Arzy, S., I. Molnar-Szakacs, and O. Blanke. "Self in Time: Imagined Self-Location Influences Neural Activity Related to Mental Time Travel." *Journal of Neuroscience* 28 (2008): 6502–6507.

Beauregard, M., and V. Paquette. "Neural Correlates of a Mystical Experience in Carmelite Nuns." *Neuroscience Letters* 405 (2006): 186–190.

Belzen, J. A., and R. W. Hood. "Methodological Issues in the Psychology of Religion: Toward Another Paradigm?" *Journal of Psychology* 140 (2006): 5–28.

Bonson, K. R., J. W. Buckholtz, and D. L. Murphy. "Chronic Administration of Serotonergic Antidepressants Attenuates the Subjective Effects of LSD in Humans." *Neuropsychopharmacology* 14 (1996): 425–436.

Borg, J., B. Andree, H. Soderstrom, and L. Farde. "The Serotonin System and Spiritual Experiences." *American Journal of Psychiatry* 160 (2003): 1965–1969.

Bucke, R. M. *Cosmic Consciousness : A Study in the Evolution of the Human Mind.* Philadelphia: Innes & Sons, 1901.

Burris, C. T. "The Mysticism Scale: Research Form D (M Scale)." In: Hill, P. C., and R. W. J. Hood, eds. *Measures of Religiosity.* Birmingham, Ala.: Religious Education Press, 1999, pp. 363–367.

Carhart-Harris, R. L., Williams, T. M. Sessa, B. Tyacke, R. J. Rich, A. S. Feilding, D. J., Nutt. "The Administration of Psilocybin to Healthy, Hallucinogen-Experienced Volunteers in AaMock-Functional Magnetic Resonance Imaging Environment: A Preliminary Investigation of Tolerability." *Journal of Psychopharmacology* (2010).

Devinsky, O., and G. Lai. "Spirituality and Religion in Epilepsy." *Epilepsy & Behavior* 12 (2008): 636–643.

Dostoyevsky, F., trans. C. Garnett. *The Idiot.* New York: Dell, 1959.

Eckhart, trans. R. B. Blakney. *Meister Eckhart, a Modern Translation.* New York: Harper & Row, 1941.

Erritzoe, D., V. G. Frokjaer, S. Haugbol, L. Marner, C. Svarer, K. Holst, W. F. Barré, et al. "Brain Serotonin 2A Receptor Binding: Relations to Body Mass Index, Tobacco and Alcohol Use." *Neuroimage* 46 (2009): 23–30.

Fay, R., and L. Kubin. "Pontomedullary Distribution of 5-HT2A Receptor-Like Protein in the Rat." *Journal of Comparative Neurology* 418 (2000): 323–345.

Fisher, P. M., C. C. Meltzer, J. C. Price, R. L. Coleman, S. K. Ziolko, C. Becker, E. L. Moses-Kolko, et al. "Medial Prefrontal Cortex 5-HT(2A) Density Is Correlated with Amygdala Reactivity, Response Habituation, and Functional Coupling." *Cerebral Cortex* 19 (2009): 2499–2507.

Frokjaer, V. G., E. L. Mortensen, F. A. Nielsen, S. Haugbol, L. H. Pinborg, K. H. Adams, C. Svarer, et al. "Frontolimbic serotonin 2A receptor binding in healthy subjects is associated with personality risk factors for affective disorder." *Biological Psychiatry* 63 (2008): 569–576.

Gouzoulis-Mayfrank, E., M. Schreckenberger, O. Sabri, C. Arning, B. Thelen, M. Spitzer, K. A. Kovar, et al. "Neurometabolic Effects of Psilocybin, 3,4-Methylenedioxyethylamphetamine (MDE) and D-Methamphetamine in Healthy Volunteers. A Double-Blind, Placebo-Controlled PET Study with [18F]FDG." *Neuropsychopharmacology* 20 (1999): 565–581.

Griffiths, R., W. Richards, M. Johnson, U. McCann, and R. Jesse R. "Mystical-Type Experiences Occasioned by Psilocybin Mediate the Attribution of Personal Meaning and Spiritual Significance 14 Months Later." *Journal of Psychopharmacology* 22 (2008): 621–632.

Griffiths, R. R., W. A. Richards, U. McCann, and R. Jesse. "Psilocybin Can Occasion Mystical-Type Experiences Having Substantial and Sustained Personal Meaning and Spiritual Significance." *Psychopharmacology (Berl)* 187 (2006): 268–283.

Gross, C., and R. Hen. "The Developmental Origins of Anxiety." *Nature Reviews Neuroscience* 5 (2004): 545–552.

Hood, R. W., N. Ghorbani, P. J. Watson, A. F. Ghramaleki, M. N. Bing, H. K. Davison, R. J. Morris, et al. "Dimensions of the Mysticsm Scale: Confirming the Three-Factor Structure in the United States and Iran." *Journal for the Scientific Study of Religion* 40 (2001): 691–705.

Hughes, J. R. "The Idiosyncratic Aspects of the Epilepsy of Fyodor Dostoevsky." *Epilepsy & Behavior* 7 (2005): 531–538.

Hurlemann, R., T. E. Schlaepfer, A. Matusch, H. Reich, N.J. Shah, K. Zilles, W. Maier, et al. "Reduced 5-HT(2A) Receptor Signaling Following Selective Bilateral Amygdala Damage." *Social Cognitive & Affective Neuroscience* 4 (2009): 79–84.

Inge, W. R. *The Philosophy of Plotinus: The Gifford Lectures at St. Andrews, 1917–1918*, 3rd ed. London, New York [etc.]: Longmans, Green and Co., 1948.

James, W., M. E. Marty. *The Varieties of Religious Experience: A Study in Human Nature.* Harmondsworth, Middlesex, England, and New York: Penguin Books, 1982.

James, W. Letter from William James to Benjamin Paul Blood, June 28, 1896, bMS Am 1092.9 (752), Houghton Library, Harvard University.

James, W. Letter from William James to Henry James, June 11, 1896, bMS Am 1092.9 (2770), Houghton Library, Harvard University.

James, W. *The Will to Believe and Other Essays in Popular Philosophy, and Human Immortality.* [New York]: Dover Publications, 1956.

Janik, V. M., L. S. Sayigh, and R. S. Wells. "Signature Whistle Shape Conveys Identity Information to Bottlenose Dolphins." *Proceedings of the National Academy of Science* 103 (2006): 8293–8297.

Jiang, X., Z. J. Zhang, S. Zhang, E. H. Gamble, M. Jia, R. J. Ursano, H. Li, et al. "5-HT2A Receptor Antagonism by MDL 11,939 During Inescapable

Stress Prevents Subsequent Exaggeration of Acoustic Startle Response and Reduced Body Weight in Rats." *Journal of Psychopharmacology* (2009).

Johnson, M., W. Richards, and R. Griffiths. "Human Hallucinogen Research: Guidelines for Safety." *Journal of Psychopharmacology* 22 (2008): 603–620.

Jung, C. G., translated by Jaffé A. *Memories, Dreams, Reflections*, rev. ed. New York: Vintage Books, 1989.

Kupers, R., V. G. Frokjaer, A. Naert, R. Christensen, E. Budtz-Joergensen, H. Kehlet, G. M. Knudsen, et al. A PET [18F]Altanserin Study of 5-HT2A Receptor Binding in the Human Brain and Responses to Painful Heat Stimulation." *Neuroimage* 44 (2009): 1001–1007.

Lilly, J. C. *The Center of the Cyclone*. New York: Julian Press, 1972.

Magalhaes, A. C., K. D. Holmes, L. B. Dale, L. Comps-Agrar, D. Lee, P. N. Yadav, L. Drysdale, et al. "CRF receptor 1 regulates anxiety behavior via sensitization of 5-HT2 receptor signaling." *Nature Neuroscience* 13 (2010): 622–629.

Meyer, J. H., A. A. Wilson, P. Rusjan, M. Clark, S. Houle, S. Woodside, J. Arrowood, et al. "Serotonin2A Receptor Binding Potential in People with Aggressive and Violent Behaviour." *Journal of Psychiatry and Neuroscince* 33 (2008): 499–508.

Monti, J. M., and H. Jantos. "Effects of Activation and Blockade of 5-HT2A/2C Receptors in the Dorsal Raphe Nucleus on Sleep and Waking in the Rat." *Progress in Neuropsychopharmacology and Biological Psychiatry* 30 (2006): 1189–1195.

Moresco, F. M., M. Dieci, A. Vita, C. Messa, C. Gobbo, L. Galli, G. Rizzo, et al. "In Vivo Serotonin 5HT(2A) Receptor Binding and Personality Traits in Healthy Subjects: A Positron Emission Tomography Study." *Neuroimage* 17 (2002): 1470–1478.

Morilak, D.A., and R. D. Ciaranello. "5-HT2 Receptor Immunoreactivity on Cholinergic Neurons of the Pontomesencephalic Tegmentum Shown by Double Immunofluorescence." *Brain Research* 627 (1993): 49–54.

Nair, D. G. "About Being BOLD." *Brain Research Reviews* 50 (2005): 229–243.

Nalivaiko, E., and A. Sgoifo. "Central 5-HT Receptors in Cardiovascular Control During Stress." *Neuroscience & Biobehavioral Reviews* 33 (2009): 95–106.

Otto, R., B. L. Bracey, and R. C. Payne. *Mysticism East and West: A Comparative Analysis of the Nature of Mysticism*. London: Macmillan and Co., Ltd., 1932.

Pinborg, L. H., H. Arfan, S. Haugbol, K. O. Kyvik, J. V. Hjelmborg, C. Svarer, V. G. Frokjaer, et al. "The 5-HT2A Receptor Binding Pattern in the Human Brain Is Strongly Genetically Determined." *Neuroimage* 40 (2008): 1175–1180.

Plotinus, A. H. Armstrong, P. Henry, and H.-R. Schwyzer. *Plotinus*. Cambridge, Mass., and London: Harvard University Press; W. Heinemann, 1966.

Plotinus, A. H. Armstrong. *Plotinus*, rev. ed. Cambridge, Mass.: Harvard University Press, 1994.

Rasmussen, H., D. Erritzo, R. Andersen, B. H. Ebdrup, B. Aggernaes, B. Oranje, J. Kalbitzer, et al. "Decreased Frontal Serotonin2a Receptor Binding in Antipsychotic-Naive Patients with First-Episode Schizophrenia." *Archives of General Psychiatry* 67 (2010): 9–16.

Reiss, D., and L. Marino. "Mirror Self-Recognition in the Bottlenose Dolphin: A Case of Cognitive Convergence." *Proceedings of the National Academy of Science* 98 (2001): 5937–5942.

Russell, B. *Mysticism and Logic*. Mineola, N.Y.: Dover Publications, 2004.

Saver, J. L., and J. Rabin. "The Neural Substrates of Religious Experience." *Journal of Neuropsychiatry & Clinical Neuroscience* 9 (1997): 498–510.

Serafetinides, E. A. "The Significance of the Temporal Lobes and of the Hemispheric Dominance in the Production of the LSD-25 Symptomatology in Man: A Study of Epileptic Patients Before and After Temporal Lobectomy." *Neuropsychologia* 3 (1965): 69–79.

Soloff, P. H., J. C. Price, N. S. Mason, C. Becker, and C. C. Meltzer. "Gender, Personality, and Serotonin-2A Receptor Binding in Healthy Subjects." *Psychiatry Research* 181 (2010): 77–84.

Stace, W. T. *Mysticism and Philosophy*. London: Macmillan, 1960.

Stace, W. T. *The Teachings of the Mystics : Being Selections from the Great Mystics and Mystical Writings of the World*. New York: New American Library, 1960.

Stockmeier, C. A. "Involvement of Serotonin in Depression: Evidence from Postmortem and Imaging Studies of Serotonin Receptors and the Serotonin Transporter." *Journal of Psychiatric Research* 37 (2003): 357–373.

Teresa, E. A. Peers. *Interior Castle*, Image ed. New York: Doubleday, 2004.

Trimble, M., and A. Freeman. "An Investigation of Religiosity and the Gastaut-Geschwind Syndrome in Patients with Temporal Lobe Epilepsy." *Epilepsy & Behavior* 9 (2006): 407–414.

Urgesi, C., S. M. Aglioti, M. Skrap, and F. Fabbro. "The Spiritual Brain: Selective Cortical Lesions Modulate Human Self-Transcendence." *Neuron* 65 (2010): 309–319.

Vaughan, R. A. *Hours with the Mystic:, A Contribution to the History of Religious Opinion*, 6th ed. New York,: Chas. Scribner's Sons, 1893.

Vollenweider, F. X., M. F. Vollenweider-Scherpenhuyzen, A. Babler, H. Vogel, and D. Hell. "Psilocybin Induces Schizophrenia-Like Psychosis in Humans via a Serotonin-2 Agonist Action." *Neuroreport* 9 (1998): 3897–3902.

Voon, V., C. Gallea, N. Hattori, M. Bruno, V. Ekanayake, and M. Hallett. "The Involuntary Nature of Conversion Disorder." *Neurology* 74 (2010): 223–228.

Weisstaub, N. V., M. Zhou, A. Lira, E. Lambe, J. González-Maeso, J. P. Hornung, E. Sibille, et al. "Cortical 5-HT2A Receptor Signaling Modulates Anxiety-Like Behaviors in Mice." *Science* 313 (2006): 536–540.

Wood, J. N., and J. Grafman. "Human Prefrontal Cortex: Processing and Representational Perspectives." *Nature Reviews Neuroscience* 4 (2003): 139–147.

Yoon, H. K, J. C. Yang, H. J. Lee, Y. K. Kim. "The association between serotonin-related gene polymorphisms and panic disorder." *Journal of Anxiety Disorders* 22 (2008): 1529–1534.

EPILOGUE

Hume, D. *Of Miracles*. La Salle, Ill.: Open Court, 1985.

Tierney, J. "Hallucinogens Have Doctors Tuning In Again." *New York Times*, April 11, 2010.

ART

Addis, D. R., M. Moscovitch, M. P. McAndrews. "Consequences of Hippocampal Damage Across the Autobiographical Memory Network in Left Temporal Lobe Epilepsy." *Brain* 130 (2007): 2327–2342. Adopted with permission.

Bradford Cannon Papers, 1923–2003, H MS c240. Harvard Medical Library, Francis A. Countway Library of Medicine Boston, Mass.

Damasio, H., T. Grabowski, R. Frank, A. M. Galaburda, A. R. Damasio. "The Return of Phineas Gage: Clues About the Brain from the Skull of a Famous Patient." Science 264 (1994): 1102–1105.

Gazzaniga, M. S., J. LeDoux *The Integrated Mind.* New York: Plenum Press, 1978.

Maquet, P., P. Ruby, A. Maudoux, G. Albouy, V. Sterpenich, T. Dang-Vu, M. Desseilles, M. Boly, F. Perrin, P. Peigneux, S. Laureys. "Human Cognition During REM Sleep and the Activity Profile Within Frontal and Parietal Cortices: A Reappraisal of Functional Neuroimaging Data." *Progress in Brain Research* 150 (2005): 219–227. Adopted with permission.

Zeman, A. "Persistent Vegetative State." *Lancet* 350 (1997): 795–799. Adopted with permission.

ACKNOWLEDGMENTS

Only after completing this book do I realize how a large team came together to make it a reality. Unfortunately space does not permit me to acknowledge all of my debts.

It is for my family that I am most grateful. My beautiful wife, Anne, with our children, Sarah, Jessica, and Matthew, stood alongside me during the many hours this book demanded.

Writing this book would have been unimaginable without the friendship of Gary Handwerk and Rogers Smith over the decades. They helped refine my thinking in crucial ways and gave me unwavering support when I needed it most.

Earl Feringa first ignited within me an excitement for clinical neurology. Under Joseph Bicknell I became a clinical neurologist, and he has kept me from straying too far from my root and center. David B. Clark taught me to see with greater acuity the people behind their neurological plights.

Among many good things, Joseph Rebar first drew my attention to John C. Lilly. Jim Clark often pointed me in the right direction.

Allan Hobson provided encouragement, as well as made clear to me the importance of subjective experience, and within it the possibilities of what REM consciousness might be.

I am grateful to my colleagues at the University of Kentucky, especially Rick Lofgren, Paul DePriest, Stephen Ryan (to whom I am particularly indebted), and Nicholas Fee, who gave me the flexibility in my day-to-day responsibilities necessary to write.

The New Mexican hospitality of Larry Johnson and Leslie Jedrey at Soaring Eagle Lodge beside the San Juan River helped tremendously to bring this book together. Jimmy Calvin and T. J. Massey guided me from turbulence to tranquility over the span in which I wrote this work.

My research collaborators Frederick Schmitt, Michelle Mattingly, and Sherman Lee were instrumental to shedding light on the mystery of near-death experience. Although we hold different beliefs about the science of near-death experiences, I thank Jeff and Jody Long for bringing research subjects to my attention.

My appreciation extends to my many colleagues who have brought cases to my attention, in particular Oliver Sacks.

Gail Ross and Howard Yoon believed in this project from its inception and reared it to fruition.

Karen Gallant crafted superb figures and drawings. Kenneth Wapner, my personal editor, contributed in ways big and small; he roused me to overcome my instincts and write from the first-person perspective. Stephen Morrow at Penguin helped immeasurably to cement this work into a coherent whole.

My intent in this work was to raise both questions and answers. As time passes, I will be found wrong on some of the answers I have offered. This is science's true beauty. All errors in this book fall squarely on my shoulders alone.

I am extremely grateful to those who took me into their confidence and shared their experiences with me. I have told only part of their stories, an omission committed with cold and deliberate intent.

The clinical stories making up this book's spiritual journey must be altered in their telling. Identities must be hidden and demographics changed. Those facts judged minor must be omitted or changed so that the flow of events can fit on these pages and be easily understood—that is convenient for you, the reader. This manipulation, no matter how subtle and well intended, brings with it some falseness. Of course all accounts of spiritual experience suffer from this illness in varying degrees. In my defense, the core of what I say remains steadfastly faithful to the larger truth—as far as I can tell—and in this I take comfort. A diamond fragmented to catch the light and someone's eye is no less a diamond than when it was found whole in its natural roughness.

INDEX

Note: Page numbers in *italics* refer to illustrations.